A Concise Introduction
to Calculus

SERIES ON UNIVERSITY MATHEMATICS

Editors:

W Y Hsiang : Department of Mathematics, University of California, Berkeley, CA 94720, USA

T T Moh : Department of Mathematics, Purdue University, W. Lafayette IN 47907, USA
e-mail: ttm@math.purdue.edu   Fax: 317-494-6318

S S Ding : Department of Mathematics, Peking University, Beijing, China

M C Kang : Department of Mathematics, National Taiwan University, Taiwan, China

M Miyanishi : Department of Mathematics, Osaka University, Toyonaka, Osaka 560, Japan

---

SERIES ON UNIVERSITY MATHEMATICS VOL. 6

A CONCISE INTRODUCTION TO
CALCULUS

W Y Hsiang

Department of Mathematics
University of California
Berkeley, USA

World Scientific
Singapore • New Jersey • London • Hong Kong

Published by

World Scientific Publishing Co. Pte. Ltd.

5 Toh Tuck Link, Singapore 596224

USA office: 27 Warren Street, Suite 401-402, Hackensack, NJ 07601

UK office: 57 Shelton Street, Covent Garden, London WC2H 9HE

Library of Congress Cataloging-in-Publication Data
Hsiang, Wu Yi, 1937–
   A concise introduction to calculus/ Wu-Yi Hsiang.
     p. cm. -- (Series on university mathematics ; vol. 6)
   ISBN-13 9789810219000 -- ISBN-10 9810219008
   ISBN-13 9789810219017 (pbk) -- ISBN-10 9810219016 (pbk)
   1. Calculus. I. Title. II. Series.
QA303.H747   1995
515--dc20   95-15402
                                 CIP

British Library Cataloguing-in-Publication Data
A catalogue record for this book is available from the British Library.

Introduction

Before one starts to undertake the task of learning "calculus", it is rather natural to inquire:

"What is calculus all about and what is it good for?"

We believe that a beginner is entitled to get an outright answer of such basic questions, if possible, at the outset.

Let us begin by observing that we are living in an ever-changing universe full of all kinds of dynamic phenomena and systems with variable quantities. For example, although we don't quite feel the motion, the earth is, in fact, spinning daily and rotating around the sun yearly; the so-called "*law of gases*" states the definitive correlation between the volume, the pressure and the temperature of a given amount of gas; the so-called "*laws of trigonometry*" provides a set of interlocking fundamental relationships among the three angles and the three sides of an arbitrary triangle. The changing of weather and the behavior of economy are still rather difficult to predict because such systems consist of too many variable parameters and the interlocking correlations among so many variables are far from being reasonably understood (or even adequately grasped). Anyway, one definitely needs a type of mathematics which provides the general framework and basic tools for *quantitative analysis of correlations among variable quantities*. This is exactly what calculus is all about and what calculus is good for! In short, calculus is the basic setting and fundamental theory of the *mathematics of variable quantities* which provides the basic tools and the general framework for analyzing all kinds of correlations among variable quantities. Here, someone may wonder:

Why not just call it "mathematics of variable quantities" rather than such a peculiar name "calculus"? Of course, there are more substantial questions about calculus other than the above question on its name. For example:

(i) What are the origin and intuitive source of calculus?
(ii) What are the basic concepts and the fundamental methodology of calculus?
(iii) What are the general framework and the basic structure of calculus?
(iv) What are the foundational theory and the typical applications of calculus?

The purpose of this concise introduction to calculus is to provide matter-of-fact, straight answers to the above set of basic questions about calculus, thus providing a good beginning for beginners in their journey of learning calculus.

Contents

CHAPTER 1

Numbers, Variables and Functions

§ 1. Real Numbers and Measurement

Numbers are the basic tools for *quantitative analysis*. In essence, numbers are *systems of mathematical symbols* designed for the purpose of concisely representing various quantitative problems and providing an efficient framework for studying and solving them. Generally speaking, there are two basic types of quantities, namely, those quantities of *counting-type* such as peoples, eggs, trees, etc., which have *individual natural units*; and those quantities of *measurement-type* such as length, area, volume, weight, time, etc., which are always *infinitely divisible* and hence impossible to have natural (indivisible) units. In daily usage, one usually distinguishes the above two types of quantities by asking "how many?" for the former, while for the latter asking "how much?". In mathematics, quantities of the counting-type are represented by *natural numbers* while quantities of the measurement-type are represented by *real numbers*. In other words, the *natural number system*, N, and the *real number system*, $\mathbb{R}$, are exactly the mathematical systems designed to handle quantitative problems of the counting-type and that of the measurement-type respectively.

The procedure of measurement is *not* as primitive as that of counting. In fact, as one shall see in the following discussion, it is quite involved both conceptually and technically. Let us take the *length of intervals* as a concrete example to analyze the concept of

1

measurement. Roughly speaking, the measurement of the length of
an interval consists of two steps, namely, (i) to choose a *unit* of length
such as a foot, a yard, a meter, a light-year, etc., and (ii) to determine
the "*ratio*" between the length of a given interval and that of the
chosen unit (the ratio is a real number). The first step is more or less
arbitrary and it does not present any theoretical problem. However,
a closer look at the second step naturally leads to a basic theoretical
problem whose fundamental importance was already recognized in the
time of Ancient Greece. When geometers of antiquity analyzed the
concept of "ratio" between the lengths of two given intervals, say a
and b, they realized that the ratio of a of b can naturally be defined
to be a rational number m/n if there exists a suitable interval c such
that $a = m \cdot c$ and $b = n \cdot c$. They called such a relationship between
two given intervals "*commensurability*". Thus they were naturally led
to pose the following basic problem of commensurability:

Problem. Whether any given pair of intervals are always commensurable or, equivalently, whether the ratio between the length of two
intervals is always a rational number.

In the daily usage of measurement of length, one uses a yardstick,
say of unit length, to perform a section-by-section matching. If the
given interval consists of exactly m sections of unit length, then the
ratio is the integer m. Otherwise, there remains a last section shorter
than the unit yardstick. One naturally tries to equally subdivide
the unit length into suitable subunits and continues to measure the
remainder by the subunit. The question here is that whether there
always exists a suitable subunit, say $1/n$ of unit, which makes the remainder exactly an integral multiple of that subunit. Thus, the above
problem naturally arises from daily practice of length measurement.
However, we shall discuss in here the deeper theoretical significance
of this problem.

If the answer to the above problem is affirmative, then the
ratio between any pair of intervals is always a rational number and
the concept of length is simple and elementary. However, if the
answer to the above problem is negative, namely, there do exist

non-commensurable pairs of intervals, then the "ratio" between such a pair is something yet to be defined and it is definitely *not* a rational number!

Notice that length is clearly the most basic geometric quantity and that other geometric quantities such as angle, area, volume, etc., can all be expressed in terms of length. Therefore, the above problem of commensurability of intervals is of fundamental importance right at the foundation of quantitative geometry. In fact, the true significance of this issue will be fully revealed only through a discussion of some major historical events concerning the above problem.

1.1. *Foundation of quantitative geometry (First attempt, mainly due to the Pythagorean school, 6th century B.C.)*

Roughly speaking, the Pythagorean school took the universal validity of commensurability as a basic "axiom" and proceeded to build a rather impressive foundation of quantitative geometry based upon such an axiom. For example, they gave the following proofs of the area formula of rectangles and the similar triangle theorems based upon such an axiom.

Area formula of rectangle. Let l and w be the length and the width of a rectangle and u be the unit of length. If $l : u = \frac{m}{n}$, $w : u = \frac{p}{q}$, then the ratio between the areas of the rectangle and the unit square is equal to $\frac{m}{n} \cdot \frac{p}{q}$.

Proof. There exist a and b such that

$$l = m \cdot a, \quad u = n \cdot a, \quad w = p \cdot b, \quad u = q \cdot b.$$

Therefore, the given rectangle can be cut into $m \cdot p$ small rectangles while the unit square can be cut into $n \cdot q$ small rectangles, all of them have a and b as their side lengths (cf. Fig. 1).

Hence the area of the rectangle in $m \cdot p$ times the area of the small one while the area of the unit square is $n \cdot p$ times the area of the small one. Thus the ratio between their areas is equal to $\frac{mp}{nq}$.   □

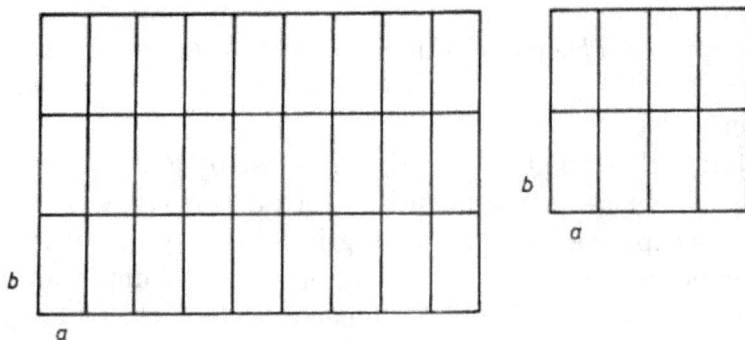

Fig. 1

Similar triangle theorem. *Let $\triangle ABC$ and $\triangle A'B'C'$ be two triangles with equal corresponding angles, namely*

$$\angle A = \angle A', \ \angle B = \angle B' \text{ and } \angle C = \angle C'.$$

Then their corresponding sides are proportionate, namely

$$\overline{AB} : \overline{A'B'} = \overline{AC} : \overline{A'C'} = \overline{BC} : \overline{B'C'}.$$

Proof. Let us first consider the special case that one pair of corresponding sides has integral ratio, say $\overline{AB} : \overline{A'B'} = n$. We shall show by induction on n that the other two pairs also have their ratios equal to n.

Subdivide $\overline{AB}$ into n equal segments by $\{B_i, \ 1 \le i \le n - 1\}$, namely

$$\overline{AB_1} = \overline{B_1 B_2} = \cdots = \overline{B_{n-1} B} = \overline{A'B'}. \tag{1}$$

Let C_{n-2}, C_{n-1} be points on AC such that

$$\overline{B_{n-2} C_{n-2}} // \overline{B_{n-1} C_{n-1}} // \overline{BC}. \tag{2}$$

Then $\triangle AB_{n-2} C_{n-2}$ and $\triangle AB_{n-1} C_{n-1}$ also have the same set of angles as that of $\triangle A'B'C'$. Therefore, by the induction assumption

$$\overline{AC_{n-2}} = (n-2)\overline{A'C'}, \ \overline{B_{n-2} C_{n-2}} = (n-2)\overline{B'C'}$$
$$\overline{AC_{n-1}} = (n-1)\overline{A'C'}, \ \overline{B_{n-1} C_{n-1}} = (n-1)\overline{B'C'}. \tag{3}$$

Choose E, D on the line of $B_{n-1}C_{n-1}$ so that

$$C_{n-2}E//DC//AB. \tag{4}$$

Then

$$\overline{C_{n-2}E} = \overline{B_{n-2}B_{n-1}} = \overline{B_{n-1}B} = \overline{DC} \tag{5}$$

and hence $\square C_{n-2}ECD$ is a parallelogram. Therefore

$$\overline{C_{n-2}C_{n-1}} = \overline{C_{n-1}C}, \ \overline{EC_{n-1}} = \overline{C_{n-1}D} \tag{6}$$

(the diagonals of a parallelogram bisect each other).

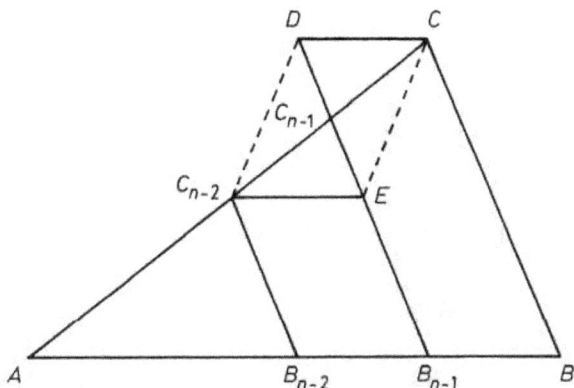

Fig. 2

Hence,

$$\begin{aligned}
\overline{AC} &= \overline{AC_{n-1}} + \overline{C_{n-1}C} = \overline{AC_{n-1}} + \overline{C_{n-2}C_{n-1}} \\
&= (n-1)\overline{A'C'} + [(n-1)\overline{A'C'} - (n-2)\overline{A'C'}] = n\overline{A'C'} \\
\overline{BC} &= \overline{B_{n-1}D} = \overline{B_{n-1}C_{n-1}} + \overline{C_{n-1}D} = \overline{B_{n-1}C_{n-1}} + \overline{EC_{n-1}} \\
&= (n-1)\overline{B'C'} + [(n-1)\overline{B'C'} - (n-2)\overline{B'C'}] = n \cdot \overline{B'C'}.
\end{aligned} \tag{7}$$

Next let us consider the case that a pair of corresponding sides has rational ratio, say $\overline{AB} : \overline{A'B'} = \frac{m}{n}$. We shall prove that the other two pairs also have their ratios equal to $\frac{m}{n}$. Choose B_1 (resp. B_1') on $\overline{AB}$ (resp. $\overline{A'B'}$) such that

$$\overline{AB_1} = \overline{A'B_1'} \text{ and } \overline{AB} = m \cdot \overline{AB_1}, \ \overline{A'B'} = n\overline{A'B_1'}.$$

Then, choose C_1 (resp. C_1') on $\overline{AC}$ (resp. $\overline{A'C'}$) such that $\overline{B_1C_1}//\overline{BC}$ (resp. $\overline{B_1'C_1'}//\overline{B'C'}$). It is easy to see that $\triangle AB_1C_1 \cong \triangle A'B_1'C_1'$ and the above proof applies to $\triangle ABC$ and $\triangle AB_1C_1$ (resp. $\triangle A'B'C'$ and $\triangle A'B_1'C_1'$). Hence

$$\overline{AC} = m\overline{AC_1}, \ \overline{BC} = m\overline{B_1C_1}$$

$$\overline{A'C'} = n\overline{A'C_1'}, \ \overline{B'C'} = n\overline{B_1'C_1'} \tag{8}$$

$$\overline{AC_1} = \overline{A'C_1'}, \ \overline{B_1C_1} = \overline{B_1'C_1'}.$$

Thus

$$\overline{AC} : \overline{A'C'} = \overline{BC} : \overline{B'C'} = \frac{m}{n}.$$

$\square$

Historical remarks

(i) At the time of Pythagorean (6th–5th century B.C.), the "universal validity" of commensurability was taken as a self-evident axiom. Therefore, the above proofs of area formula and the similar triangle theorem are regarded as perfectly general and complete.

(ii) The Pythagoras Theorem and the above similar triangle theorem are the two fundamental theorems of quantitative geometry, moreover, the proof of Pythagoras Theorem is necessarily based upon the above area formula. Therefore, the area formula and the similar triangle theorem are truly of fundamental importance.

(iii) Unfortunately, the basic "axiom" of universal validity of commensurability is, in fact, *not* true. The *existence of non-commensurable* pairs of intervals was first discovered by Hippasus, a disciple of Pythagoras. Therefore, the above proofs are *only* the proofs of the *commensurable case* and hence theoretically *incomplete!*

Let us first discuss the discovery of Hippasus.

1.2. *Discovery and Proof of existence of non-commensurable pairs of intervals*

First of all, notice that the commensurability problem is a *purely theoretical* problem whose significance and importance lie in the foundation of geometry. (For practical purposes, any two intervals can be regarded as commensurable simply by omitting a very, very small, i.e., practically nil, remainder.) Therefore, the existence of non-commensurable pairs of intervals can only be demonstrated by theoretical proofs. Let us begin with a theoretical criterion of commensurability.

A criterion of commensurability. Suppose that a and b are a pair of commensurable intervals, namely, there exists a common "yardstick" c such that both a and b are integral multiples of c, say $a = m \cdot c$ and $b = n \cdot c$. Let $l = (m, n)$ be the greatest common divisor of m and n. Then $c' = l \cdot c$ is clearly the longest common yardstick of a and b. Corresponding to the Euclid algorithm of computing l from m and n, one has the following "*geometric algorithm*" of finding the longest common yardstick c' from a and b, namely:

Use the shorter one, say b, as the yardstick to measure the longer one, say a. If a is an integral multiple of b, then b itself is the longest common yardstick. Otherwise, one has a remainder r_1 shorter than b. Next use r_1 as the yardstick to measure b. If b is an integral multiple of r_1, then r_1 is the longest common yardstick. Otherwise, one has another remainder r_2 shorter than r_1. Keep going until the last remainder, say r_k, can integrally measure the preceding remainder r_{k-1} (i.e., without remainder). Then r_k is the longest common yardstick of a and b that we are seeking.

On the other hand, if one can *prove* that the above algorithm will never end for a specific, given pair of intervals, then such a pair is proven to be *non-commensurable*. This was exactly how Hippasus proved the non-commensurability of some specific pairs of intervals, the first one discovered by him is the following, namely

Example 1. Let a and b be the diagonal and the side of a regular pentagon. Then a and b are non-commensurable.

Proof. The proof is to show that the algorithm of alternating measurement, applying to such a pair $\{a, b\}$ will never end!

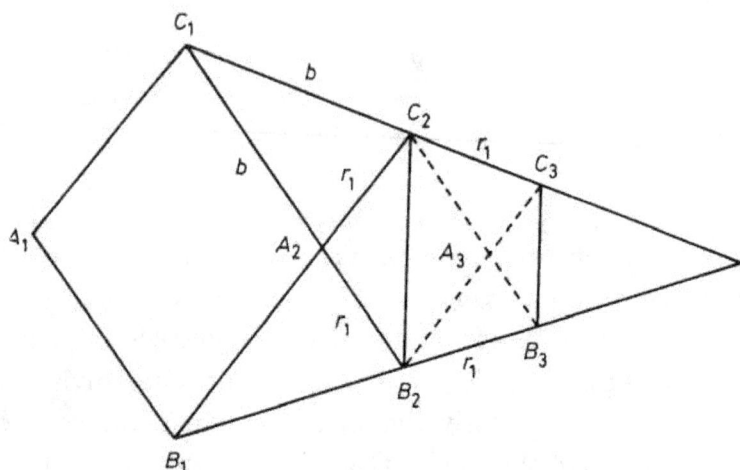

Fig. 3

As indicated in Fig. 3, $A_1B_1B_2C_2C_1$ is a regular pentagon whose side length and diagonal length are b and a respectively and its five inner angles are all equal to $\frac{3\pi}{5}$. $\triangle C_1B_2C_2$ is an isosceles triangle, thus

$$\angle C_1B_2C_2 = \angle B_2C_1C_2 = \frac{1}{2}\left(\pi - \frac{3\pi}{5}\right) = \frac{\pi}{5} \tag{9}$$

and the same reason shows that $\angle B_2C_2B_1 = \frac{\pi}{5}$. Therefore $\triangle A_2B_2C_2$ is also an isosceles triangle and $\angle B_2A_2C_2 = \pi - \frac{\pi}{5} - \frac{\pi}{5} = \frac{3\pi}{5}$. Moreover,

$$\angle C_1A_2C_2 = \pi - \frac{3\pi}{5} = \frac{2\pi}{5}, \ \angle C_1C_2A_2 = \frac{3\pi}{5} - \frac{\pi}{5} = \frac{2\pi}{5} \tag{10}$$

and hence $\triangle C_1A_2C_2$ is also an isosceles triangle.

Thus

$$a = \overline{C_1B_2} = \overline{C_1A_2} + \overline{A_2B_2} = b + r_1, \ r_1 = \overline{A_2B_2} = \overline{A_2C_2}. \tag{11}$$

Extend $\overline{B_1B_2}$ (resp. $\overline{C_1C_2}$) to B_3 (resp. C_3) such that $\overline{B_2B_3} = \overline{C_2C_3} = r_1$. Then, as indicated in Fig. 3, the pentagon $A_2B_2B_3C_3C_2$ is again a regular one! (The proof of this fact is a simple exercise.) Moreover, its diagonal length is b while its side length is r_1. Therefore, as one proceeds to measure b by r_1 as the yardstick, the geometric situation is exactly the same as before, namely, the remainder is the difference between the diagonal and the side of a regular pentagon. Thus

$$b = r_1 + r_2, r_1 = r_2 + r_3, \ldots, r_{k-1} = r_k + r_{k+1}, \ldots \qquad (12)$$

where the pair $\{r_{k-1}, r_k\}$ are always the diagonal and the side of a regular pentagon! Of course, this algorithm can never end, although the size of the k-th regular pentagon gets smaller and smaller. This proves that $\{a, b\}$ are *non-commensurable*!

Example 2. After he discovered the above astonishing example of non-commensurable pair of intervals by the above simple ingenious proof, Hippasus naturally proceeded to analyze the commensurability problem between the diagonal and the side of a square, say a' and b'. As indicated by Fig. 4, it is not difficult to show that the algebraic relationships among the remainders of the algorithm of alternating measurements are as follows, namely

$$a' = b' + r_1, b' = 2r_1 + r_2, r_1 = 2r_2 + r_3, \ldots, r_{k-1} = 2r_k + r_{k+1}, \ldots \quad (13)$$

Therefore, the geometric situations from the second one onward are all the same and hence this algorithm can never end! Thus $\{a', b'\}$ is again a non-commensurable pair.

[We leave the geometric proof of (13) as an exercise.]

Historical remarks

(i) The above discovery of non-commensurable pairs by Hippasus is a monumental milestone in the entire human civilization of rational mind. However, to his fellow Pythagoreans and contemporary

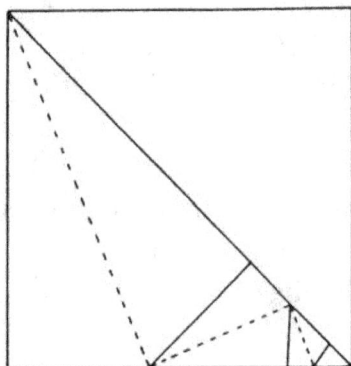

Fig. 4

geometers, this was a gigantic "*geoquake*" which rocked the whole foundation of quantitative geometry. The proofs of the area formula of rectangle and the similar triangle theorem that they prided were *no longer* complete proofs covering full generality, but rather, they were merely proofs for the *special commensurable case* only.

(ii) The historical record of this great event is unfortunately lost. However, according to some indirect sources, the following story may roughly serve as an account of what was happening to Hippasus and his great discovery: The initial reaction of his fellow Pythagoreans were shock and denial and, in order to avoid the unbearable embarrassment of public disgrace, they decided to cover it up and vowed to keep it as a secret. However, such a covering up of fundamental truth, eventually, became unbearable for the scholar Hippasus and he somehow leaked the truth of his great discovery to the outsiders (which, by the way, were often referred by the Pythagoreans simply as "the unworthies"). This made his fellow Pythagoreans furious and they condemned him to death! Naturally, he fled away. But unfortunately, the Pythagoreans were eventually able to track him down on a merchant ship in the Mediterranean and they pushed him overboard. Thus, a great hero of human civilization died for the truth. One might add here that the above story should probably be labelled as "*the first Pentagon Paper*".

(iii) Actually, his fellow Pythagoreans should be proud of such a monumental discovery by their school. Moreover, although their first attempt in building a foundation of quantitative geometry was not as perfect as they thought, it was still a major step forward and an impressive achievement by itself. Therefore, the proper reaction should be to celebrate the new discovery of their colleague, admitting the inadequacy of their proofs based upon the false axiom of universal validity of commensurability and then resolved to work for the proofs of the remaining non-commensurable case. Of course, such proofs were by no means easy to find and they naturally became the major challenge to the entire community of Greek geometers of that time. The task of rebuilding a solid foundation of quantitative geometry was finally succeeded by Eudoxus (408–355 B.C.) and his successful story is naturally our next topic of discussion.

1.3. *Eudoxian principle, the origin of the methodology of approximation*

Let us begin with some analysis of the task that Eudoxus and his contemporary were facing.

Analysis

1. In the case that two intervals a and b are *commensurable*, the ratio between their lengths has a clear simple meaning and it is a rational number. However, in the case that two intervals a and b are *non-commensurable* (such as the case of Examples 1 and 2), the meaning of the ratio between them is something yet to be defined and it is definitely *not* a rational number.

2. Although the meaning of the ratio between two non-commensurable pairs of intervals a and b is still undefined, the meaning of inequality between such a yet to be defined ratio and a given rational number $\frac{m}{n}$ such as

$$a : b > \frac{m}{n} \text{ or } a : b < \frac{m}{n}$$

is, in fact, quite clear, namely

(i) $a : b > \frac{m}{n}$ if $n \cdot a$ is longer than $m \cdot b$,

(ii) $a : b < \frac{m}{n}$ if $n \cdot a$ is shorter than $m \cdot b$.

3. Suppose that a and b are a given pair of non-commensurable intervals. Then, to a given natural number n, one may first subdivide b into n equal parts and then use $\frac{1}{n} \cdot b$ as the yardstick to measure a, thus obtaining an m such that $m \cdot \frac{1}{n}b$ is shorter than a while $(m+1)\frac{1}{n}b$ is longer than a, namely

$$\frac{m}{n} < a : b < \frac{m+1}{n}. \tag{14}$$

Therefore, the difference between $a : b$ and $\frac{m}{n}$ (resp. $\frac{m+1}{n}$) is, of course, less than $\frac{1}{n}$, although the meaning of $a : b$ is yet to be defined. By choosing n sufficiently large, the above difference can be *as small as one wishes!*

4. Suppose that $\{a, b\}$ and $\{a', b'\}$ are two given pairs of non-commensurable intervals. How do we compare their ratios $a : b$ and $a' : b'$? Suppose that $a : b < a' : b'$. Then one may choose n sufficiently large such that $\frac{1}{n}$ is smaller than the difference between the above two ratios (whatever the meaning of the difference of such two ratios may eventually be defined to be). Let m be such an integer that $\frac{m}{n} < a : b < \frac{m+1}{n}$. Then $\frac{m+1}{n}$ must be smaller than $a' : b'$, namely

$$a : b < \frac{m+1}{n} < a' : b'. \tag{15}$$

The above analysis naturally leads to the following definition enunciated by Eudoxus, namely

Eudoxian principle. Let $\{a, b\}$ and $\{a', b'\}$ be two pairs of non-commensurable intervals. If there exists a fraction $\frac{m}{n}$ such that

$$a : b < \frac{m}{n} < a' : b' \ (\text{resp. } a : b > \frac{m}{n} > a' : b')$$

then $a' : b'$ is *defined* to be larger (resp. smaller) than $a : b$. On the other hand, if any rational $\frac{m}{n}$ which is larger (resp. smaller) than

$a : b$ is also larger (resp. smaller) than $a' : b'$, then $a' : b'$ is *defined* to be equal to $a : b$, namely, a necessary and sufficient condition of $a : b = a' : b'$ is that

$$na \left\{ {> \atop <} \right\} m \cdot b \Leftrightarrow na' \left\{ {> \atop <} \right\} mb' \qquad (16)$$

for all m and n.

The above *criterion* of the equality between the ratios of two pairs of non-commensurable intervals is undoubtedly correct. However, such a criterion needs to verify that $a : b$ and $a' : b'$ have *identical inequality relationships with all rational numbers* in order to establish the *equality* between them. That means one has to check *infinitely many inequalities* in order to obtain a single equality. One cannot help but wonder about the usefulness of such a criterion. The first major application of the above Eudoxian principle is to provide a firm foundation of geometry. The following proof of the similar triangle theorem for the remaining case of non-commensurable ones is a typical example of such an application.

Example 3. A proof of similar triangle theorem for the non-commensurable case by Eudoxian principle. Let $\frac{m}{n}$ be an arbitrary fraction number. We shall show that

$$\overline{AB} : \overline{A'B'} \left\{ {> \atop <} \right\} \frac{m}{n} \Rightarrow \overline{AC} : \overline{A'C'} \text{ and } \overline{BC} : \overline{B'C'} \left\{ {> \atop <} \right\} \frac{m}{n}. \qquad (17)$$

Set B_1 to be the point on AB such that $AB_1 : \overline{A'B'} = \frac{m}{n}$, and C_1 to be the point on AC such that $B_1C_1 // BC$. Then

$$\angle B_1 = \angle B = \angle B', \ \angle C_1 = \angle C = \angle C'. \qquad (18)$$

Therefore, by the proven commensurable case of the theorem

$$\overline{AC_1} : \overline{A'C'} = \overline{B_1C_1} : \overline{B'C'} = \frac{m}{n}. \qquad (19)$$

If $\overline{AB} : \overline{A'B'} > \frac{m}{n}$ (resp. $< \frac{m}{n}$), then

$$\overline{AB} > \overline{AB_1}, \ \overline{AC} > \overline{AC_1} \text{ and } \overline{BC} > \overline{B_1C_1}$$
$$(\text{resp. } \overline{AB} < \overline{AB_1}, \ \overline{AC} < \overline{AC_1} \text{ and } \overline{BC} < \overline{B_1C_1})$$

and hence

$$\overline{AC} : \overline{A'C'} > \overline{AC_1} : \overline{A'C'} = \frac{m}{n}$$

$$\text{and} \quad \overline{BC} : \overline{B'C'} > \overline{B_1C_1} : \overline{B'C'} = \frac{m}{n}$$

$$\left(\text{resp. } \overline{AC} : \overline{A'C'} < \overline{AC_1} : \overline{A'C'} = \frac{m}{n}, \right.$$

$$\left. \text{and} \quad \overline{BC} : \overline{B'C'} < \overline{B_1C_1} : \overline{B'C'} = \frac{m}{n} \right). \tag{20}$$

This proves that

$$\overline{AB} : \overline{A'B'} = \overline{AC} : \overline{A'C'} = \overline{BC} : \overline{B'C'}. \tag{21}$$

$\square$

The above remarkable, simple proof based upon a direct application of the Eudoxian principle enables us to deduce the general case of the similar triangle theorem from the proven special case of a commensurable pair $\overline{AB}$ and $\overline{A'B'}$. Next let us apply it to analyze the area formula of rectangles.

Analysis

1. In the proven commensurable case, the area formula can be written as follows, namely

$$\square(l, w) : \square(u, u) = (l : u) \cdot (w : u) \tag{22}$$

where $\square(l, w)$ denotes the area of a rectangle with length and width equal to l and w respective and u is the chosen unit of length, thus $\square(u, u)$ is the area of the square of unit side-length (the natural unit of area).

2. In the general case that $\{l, u\}$ or $\{w, u\}$ are non-commensurable, their ratios are *no longer* rational numbers (such ratios are, nowadays, called *irrational numbers*). In fact, the meaning of the product of two irrationals (or one irrational and one rational) is something yet to be defined. Therefore, we need to clarify the exact meaning of such a multiplication among irrationals before we proceed to prove (22) for the general case with irrational $l : u$ and/or $w : u$.

3. Suppose that $\frac{m}{n}$, $\frac{p}{q}$ (resp. $\frac{m'}{n'}$, $\frac{p'}{q'}$) are fractions such that

$$\frac{m}{n} < l : u < \frac{p}{q} \quad \left(\text{resp. } \frac{m'}{n'} < w : u < \frac{p'}{q'} \right). \tag{23}$$

Then the product of $l : u$ and $w : u$ must be a number lying between the products of rationals, namely

$$\frac{m}{n} \cdot \frac{m'}{n'} < (l : u) \cdot (w : u) < \frac{p}{q} \cdot \frac{p'}{q'}. \tag{24}$$

In particular, suppose that n is a large integer and

$$\frac{m}{n} < l : u < \frac{m+1}{n} \quad \text{and} \quad \frac{m'}{n} < w : u < \frac{m'+1}{n}. \tag{23'}$$

Then

$$\frac{mm'}{n^2} < (l : u)(w : u) < \frac{(m+1)(m'+1)}{n^2}. \tag{24'}$$

Notice that the difference between the above two bounds, namely

$$\frac{(m+1)(m'+1)}{n^2} - \frac{mm'}{n^2} = \frac{m+m'+1}{n^2} = \frac{1}{n} \left(\frac{m}{n} + \frac{m'}{n} + \frac{1}{n} \right) \tag{25}$$

which becomes arbitrarily small as n getting sufficiently large.

Definition. $(l : u) \cdot (w : u)$ is the unique number lying between $\frac{m}{n} \cdot \frac{m'}{n'}$ and $\frac{p}{q} \cdot \frac{p'}{q'}$, for any

$$\frac{m}{n} < l : u < \frac{p}{q} \quad \text{and} \quad \frac{m'}{n'} < w : u < \frac{p'}{q'}.$$

[The above condition uniquely characterizes a number because the difference of (25) can be arbitrarily small.]

Proof of the area formula of rectangles (i.e., (22)) for the general case. Let $\square ABCD$ be a rectangle with $\overline{AB} = l$ and $\overline{AD} = w$ and

$$\frac{m}{n} < l : u < \frac{p}{q}, \ \frac{m'}{n'} < w : u < \frac{p'}{q'}.$$

Set B_1 and B_2 (resp. D_1 and D_2) to be points on AB (resp. AD) such that

$$\overline{AB_1} = \frac{m}{n}u, \ \overline{AB_2} = \frac{p}{q}u$$

$$\overline{AD_1} = \frac{m'}{n'}u, \ \overline{AD_2} = \frac{p'}{q'}u.$$

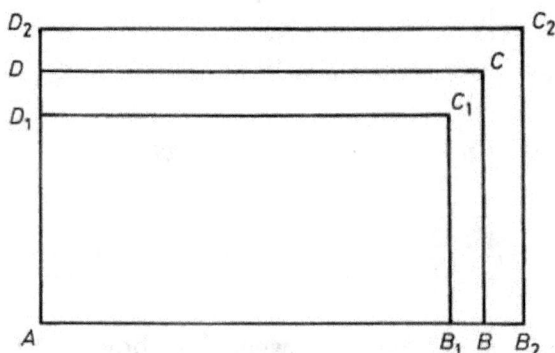

Fig. 5

Then, as indicated in Fig. 5, $\square AB_1C_1D_1$ is a rectangle ccontained in $\square ABCD$ while $\square AB_2C_2D_2$ is a rectangle containing $\square ABCD$. Moreover, it follows from the proven commensurable case

$$\square AB_1C_1D_1 : \square(u, u) = \frac{m}{n} \cdot \frac{m'}{n'}$$

$$\square AB_2C_2D_2 : \square(u, u) = \frac{p}{q} \cdot \frac{p'}{q'}.$$

Therefore

$$\frac{m}{n} \cdot \frac{m'}{n'} < \Box ABCD : \Box(u, u) < \frac{p}{q} \cdot \frac{p'}{q'}$$

and hence, by the above definition of multiplication among irrationals,

$$\Box ABCD : \Box(u, u) = (l : u) \cdot (w : u).$$

Historical remarks

(i) The Eudoxian principle provides a mathematical definition of irrationals via the comparison between a given ratio of non-commensurable pair with all rational numbers, thus enabling us to extend the proofs of the basic theorems and basic formulas of quantitative geometry from the special commensurable case to the case of full generality.

(ii) The underlying idea of Eudoxian principle is to *approximate* irrationals by rationals. It was the first monumental success of *approximation methodology* and it is the origin of modern *limit concept*, cf. Sec. 1, Chapter 3.

(iii) The real number system is the mathematical system for quantities of measurement type such as length, area, volume, weight, density, etc. It is the very foundation of quantitative analysis. Since quantities of measurement type are always divisible, it is obvious that the real number system contains the rational number system. The earlier Pythagoreans mistakenly believed that the rational number system already suffices for measurement of lengths and proceeded to build a foundational theory of quantitative geometry. Hippasus' discovery of non-commensurable pairs of intervals demonstrates the existence of irrational numbers which can be compared as the discovery of a new continent in mathematics. Eudoxian principle of approximation teaches us how to understand this new continent.

Exercises

1. Use the area formulas of triangles and rectangles to prove the Pythagoras Theorem, namely

$$a^2 + b^2 = c^2$$

   for the three side-lengths of a right-angled triangle.
   [Hint: To compute the area of the following square of $(a + b)$ in two different ways as indicated in Fig. 6.]

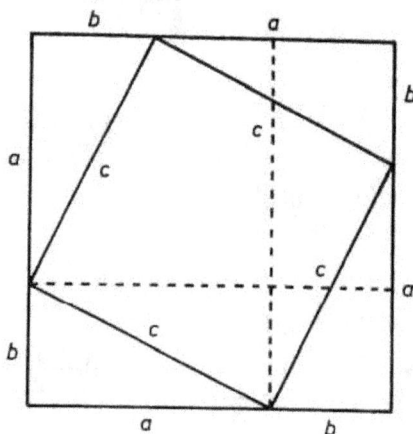

Fig. 6

2. Use the area formula of triangles and the cutting of Fig. 7 to deduce the similar triangle theorem, namely

$$\left(\frac{a}{a'}\right)^2 = \left(\frac{b}{b'}\right)^2 = \left(\frac{c}{c'}\right)^2 = \frac{\Delta}{\Delta'}$$

   where $\{a, b, c\}$, $\{a', b', c'\}$ are the side-lengths of $\triangle ABC$ and $\triangle AB'C'$ and Δ, Δ' are their areas, cf. Fig. 7.

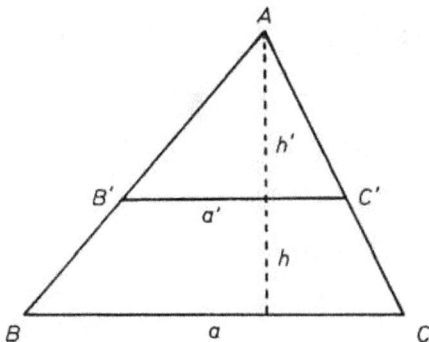

Fig. 7

3. Let a, b be the lengths of the diagonal and the side of a regular pentagon. Show that

$$a : b = \frac{1}{2}(1 + \sqrt{5}).$$

[Hint: $\triangle C_1 B_2 C_2 \sim \triangle C_2 A_2 B_2$, therefore $a : b = b : (a - b)$.]

4. Show that no squares of rationals can be equal to 2.
   [Hint: To prove by contradiction starting with $m^2 = 2n^2$ and at least one of m and n is an odd integer.]

5. Use the fact $a : b = \frac{1}{2}(1 + \sqrt{5})$ to give a construction by rule and compass for the regular pentagon with a given side-length b.

6. Use the similar triangle theorem to deduce the Pythagoras Theorem.
   [Hint: The perpendicular line CD cut the right-angled triangle $\triangle ABC$ into two triangles similar to itself.]

7. Try to express $\frac{8}{11}, \frac{7}{13}, \frac{13}{7}$ by decimals.
   [Hint: They are infinite decimals with cyclic phenomena.]

8. Analyzing the experience of Exercise 7, can you make a guess whether the decimal of an arbitrary fraction number $\frac{p}{q}$ is necessarily cyclic?

9. Can you roughly sketch a reason for your "conjecture" on the above problem.

10. Find the limit values of the following cyclic decimals, namely

$$0.010101\cdots =?$$
$$0.001001001\cdots =?$$
$$0.000100010001 =?$$
$$0.123123123 =?$$
$$0.123412341234 =?$$

11. Analyzing the experience of Exercise 10, can you make a guess whether the limit value of an arbitrary cyclic decimal is necessarily a rational number?

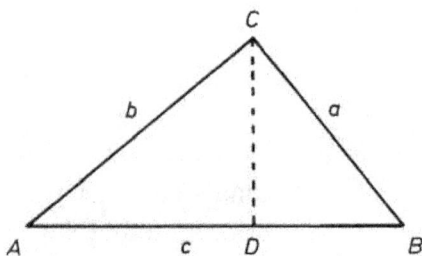

§ 2. Variables and Functions

In the mathematical analysis of a given system with variable quantities, it is convenient to represent the value of each identified variable quantity by a symbol say x, y or t, and to describe those interlocking correlations among the variable quantities of such a given system as functional relations amongst the representing symbols which are called "variables". In other words, *variables* are simply mathematical symbols that one uses to represent the values of variable quantities and *functions* are the kind of mathematical relations that one uses to describe or to investigate the *correlations* amongst variable quantities of a given system or dynamic phenomenon. Let us first mention some concrete examples before we proceed to give formal definitions of variables and functions.

Example 1. For a given elastic spring, the Hooke's law of elasticity asserts that the deformation is proportionate to the force applied. If one uses x and f to denote the magnitudes of the deformation and the force respectively then the correlation between them can be simply represented by

$$x = k \cdot f$$

where k is the elastic constant of the given spring.

Example 2. One of the simple dynamic systems of daily use is the constant speed circular motion. For example, if one chooses the center of the circle as the origin, the radius of the circle as the unit of length, then circular motion of *unit* speed can be described by a pair of basic functions, namely

$$\begin{cases} x = \cos t \\ y = \sin t \end{cases}$$

which records the changing of (x, y)-coordinate of the moving point as a pair of functions of the time t (cf. Fig. 8).

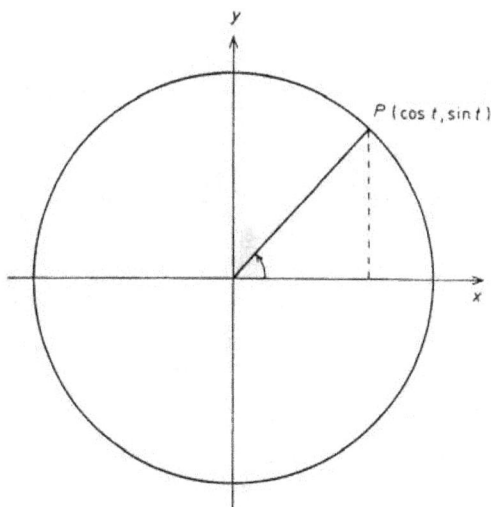

Fig. 8

Example 3. The two acute angles $\angle A$ and $\angle B$ and the three sides of a right-angled triangle $\triangle ABC$ are linked by the following functional relations, namely

$$\begin{cases} c^2 = a^2 + b^2 \text{ (Pythagoras Theorem)} \\ a = c \cdot \sin A, \ b = c \cos A \\ A + B = \frac{\pi}{2}. \end{cases}$$

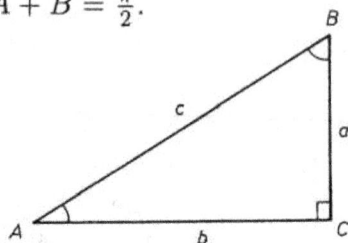

Fig. 9

Example 4. For general triangles, let A, B, C be its three angles and a, b, c be its three sides (cf. Fig. 8). Then they are related by the following basic relations, namely

$$A + B + C = \pi$$

$$\frac{\sin A}{a} = \frac{\sin B}{b} = \frac{\sin C}{c} \text{ (sine law)}$$

$$\begin{cases} \cos A = \frac{b^2 + c^2 - a^2}{2bc} \\ \cos B = \frac{a^2 + c^2 - b^2}{2ac} \text{ (cosine law)} \\ \cos C = \frac{a^2 + b^2 - c^2}{2ab}. \end{cases}$$

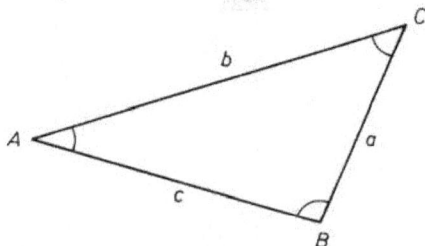

Fig. 10

Example 5. For a given amount of gas, let P, V and T be the variables representing its pressure, volume and temperature (in degree Celsius) respectively. Then the law of gases asserts that

$$PV = c(T + 273.16)$$

where c is a constant that depends on the amount of gas.

Example 6 (Postage). If one goes to a post office to send some mail, one finds that the postage is determined by the type of mail and the *weight*. Mathematically speaking, there is a postage-function defined for each type of mail. However, they all have a simple common character, namely, the acceptable weight ranges are divided into subintervals such that the postage remains a fixed constant for weights within each subinterval. We shall call this type of function *piecewise constant functions* or *step functions*.

Example 7 (Income tax and income tax rate). The *"income tax"* is a function of *"taxable income"*. However, the actual computation of income tax in terms of the taxable income involves another function called the *income tax rate* which is again a piecewise constant function. It is interesting to note the relationship between the graph of the income tax function and the income tax rate function. The former is piecewise linear while the latter is piecewise constant which is exactly the slope of the former.

Of course, there are abundant concrete examples of variable quantities and functional relations among variable quantities. The formal definitions of variables and functions are simply the mathematical setting which are abstractions of those concrete examples.

Definition. A *variable* is a symbol, such as x, y or t, which represents an arbitrary element of a domain of numbers, D, called the *domain of the given variable*.

Definition. If two given variables x and y are related in such a way that the corresponding value of y is *uniquely determined* by the value of x, then y is said to be a function of x. More generally, if a collection of variables $x_1, x_2, \ldots, x_n$ and y are related in such a way that the corresponding value of y is *uniquely determined* when the value of each x_i, $1 \leq i \leq n$, is given, then y is said to be a function of the n variables x_i, $1 \leq i \leq n$.

Examples

1. *Polynomial functions* such as

$$y = x^2, \; y = x^3 + \sqrt{2}x^2 - 3$$
$$y = (x_1 + x_2)^n, \; y = x_1^2 + x_2^2 + \cdots + x_n^2$$

are a simple basic family of functions.

2. *Trigonometric functions* such as

$$y = \sin x, \; y = \cos x, \; y = \tan x$$
$$y = \sin nx, \; y = \cos nx, \text{ etc.}$$

are a basic family of periodic functions.

3. *Piecewise constant functions* (also called step functions), namely, such a function $y = f(x)$ is defined over a union of *open intervals*

$$(a_0, a_1) \cup (a_1, a_2) \cup \cdots \cup (a_{i-1}, a_i) \cup \cdots \cup (a_{l-1}, a_l)$$

such that the restriction of $y = f(x)$ to each open interval $(a_{i-1}, a_i) = \{x \mid a_{i-1} < x < a_i\}$ takes a fixed constant value, say c_i. Because the graph of such a piecewise constant function consists of a sequence of horizontal steps of various heights, it is often called a *step-function*.

4. *Piecewise linear functions.* If $y = mx + k$, then the graph of such a function is a straight line with m as its slope. Such functions are called linear functions. Slightly more general, if $y = f(x)$ is a function defined over each closed subinterval of a suitable subdivision by a linear function, then it is called a piecewise linear function. The

graph of such a piecewise linear function consists of a connecting sequence of straight segments (cf. Fig. 12).

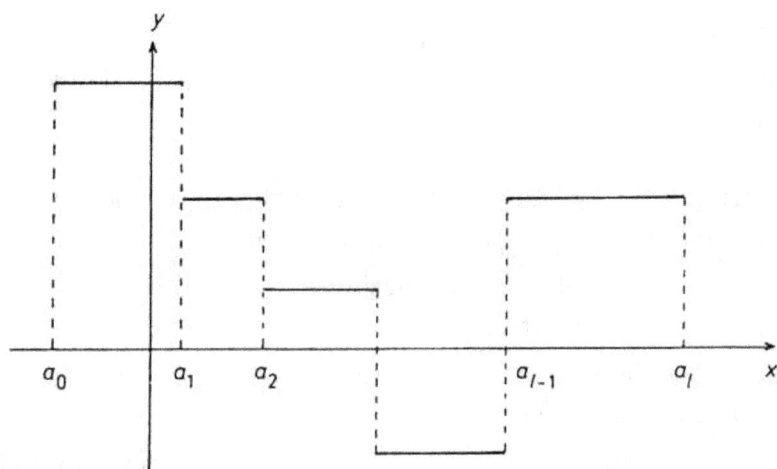

Fig. 11

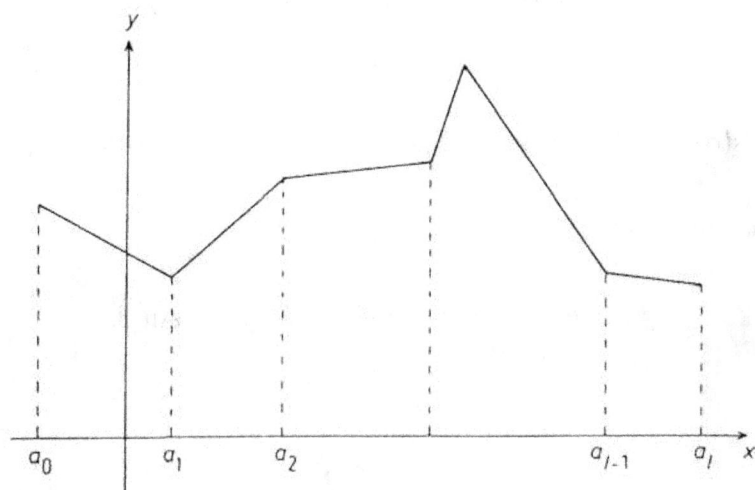

Fig. 12

Exercises

Polynomial functions and trigonometric functions are amongst the simplest and also the most basic types of functions. They are also amongst the most useful ones.

1. Determine the polynomial function of degree 1 whose values at $x = 1$ and 3 are equal to -1 and 5 respectively.

2. Determine the polynomial function of degree 2 whose values at $x = 0, 1, 2$ are equal to $0, 0, 1$ respectively.

3. Determine the polynomial function of degree 3 whose values at $x = 0, 1, 2, 3$ are equal to $0, 0, 1, 5$ respectively.

4. Let $(\cos \alpha, \sin \alpha)$ and $(\cos \beta, \sin \beta)$ be two points on the unit circle. Show that
$$(\cos \alpha - \cos \beta)^2 + (\sin \alpha - \sin \beta)^2$$
only depends on the value of $(\alpha - \beta)$. In particular
$$(\cos(\alpha-\beta)-1)^2+(\sin(\alpha-\beta)-0)^2 = (\cos \alpha-\cos \beta)^2+(\sin \alpha-\sin \beta)^2.$$

5. Use the result of 4 to show that
$$\cos(\alpha - \beta) = \cos \alpha \cos \beta + \sin \alpha \sin \beta.$$

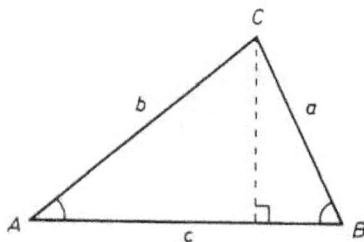

Fig. 13

6. Show that the area of $\triangle ABC$ is equal to

$$\frac{1}{2}bc \sin A.$$

7. Use 6 to prove the sine law

$$\frac{\sin A}{a} = \frac{\sin B}{b} = \frac{\sin C}{c} \left(= \frac{2\Delta}{abc} \right).$$

8. Show that

$$\begin{cases} a \cos B + b \cos A = c \\ b \cos C + c \cos B = a \\ a \cos C + c \cos A = b. \end{cases}$$

9. Use 8 to prove that

$$\begin{cases} \cos A = \frac{b^2 + c^2 - a^2}{2bc} \\ \cos B = \frac{a^2 + c^2 - b^2}{2ac} \\ \cos C = \frac{a^2 + b^2 - c^2}{2ab}. \end{cases}$$

10. Use Ex. 7 and 9 to show that

$$16\Delta^2 = (a + b + c)(a + b - c)(a - b + c)(-a + b + c).$$

11. Let R be the circumradius of $\triangle ABC$. Show that

$$\frac{\sin A}{a} = \frac{\sin B}{b} = \frac{\sin C}{c} = \frac{1}{2R}.$$

[Hint: Fig. 14.]

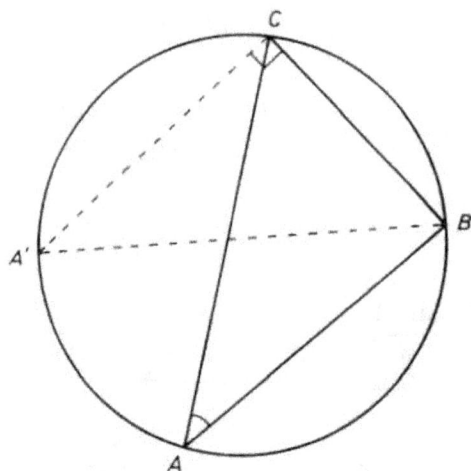

Fig. 14

12. Let r be the radius of the inscribing circle of $\triangle ABC$ and set
    $s = \frac{1}{2}(a + b + c)$. Show that
    (i) $\Delta = \frac{1}{2}r(a + b + c) = rs$
    (ii) $\tan \frac{A}{2} = \frac{r}{(s-a)}$.
    [Hint: Fig. 15.]

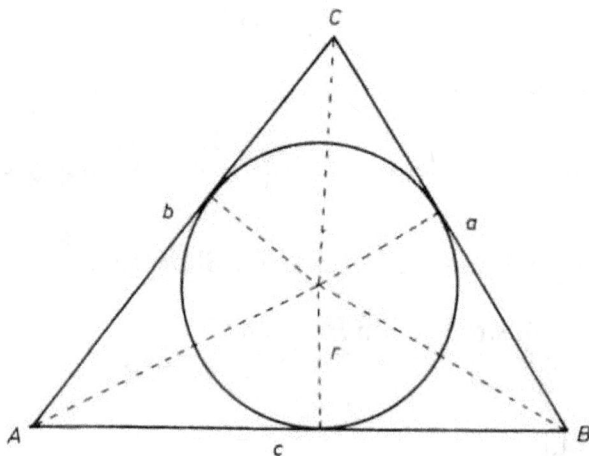

Fig. 15

13. Show that $\tan \frac{A}{2} = \frac{\sin A}{1+\cos A}$.

14. Set $\tan \frac{A}{2} = m$. Then

$$\sin A = \frac{2m}{1+m^2} \quad \cos A = \frac{1-m^2}{1+m^2}.$$

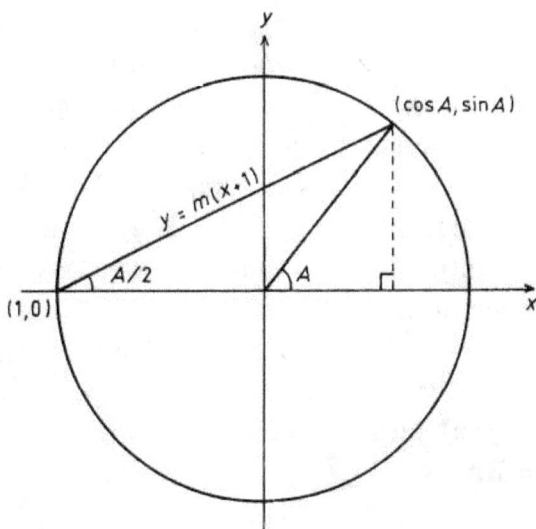

Fig. 16

[Hint: Fig. 16.]

CHAPTER 2

Basic Properties of Functions

Recall that calculus is a branch of mathematics which provides the basic techniques and general framework for *quantitative analysis* of the correlations among variable quantities of a given system or phenomenon. Thus, functions are exactly the primary subject of study in calculus. In this chapter, we shall begin our study of functions by analyzing the intuitive background of some basic features as well as some fundamental properties of functions. To avoid unnecessary complication at the very beginning, we shall only consider the simplest case of real value functions of a single real variable, namely, functions of the type $y = f(x)$ with x, y varies over certain subsets of the real number system $\mathbb{R}$.

§ 1. Monotonicity and Continuity of Functions

Monotonicity and continuity are the two basic *qualitative* properties of functions.

1.1. *Monotonicity and local extremals*

Definition. A function $y = f(x)$ is said to be *monotonically increasing* (resp. *decreasing*) over the interval $[a, b]$ (or (a, b)) if

$$x_1 < x_2 \Rightarrow f(x_1) \le f(x_2) \text{ (resp. } f(x_1) \ge f(x_2))$$

for any pair x_1, x_2 in the interval.

A function $y = f(x)$ is said to be *piecewise monotonic* if the whole interval can be suitably subdivided into finitely many subintervals such that the restriction of $y = f(x)$ to each subinterval is either monotonically increasing or monotonically decreasing.

Examples

1. A piecewise linear function is always piecewise monotonic.
2. $y = x^2$ is monotonically increasing (resp. decreasing) over $[0, a]$ (resp. $[b, 0]$), thus it is piecewise monotonic.
3. $y = x^3$ is monotonically increasing over any interval.
4. $y = x^2 + 4x + 7 = (x + 2)^2 + 3$. Hence, it is easy to see that it is monotonically increasing (resp. decreasing) over $[-2, a]$ (resp. $[b, -2]$).
5. $x = \cos t$ and $y = \sin t$ of Example 2 in Sec. 1.2 are both piecewise monotonic.

Definition. A point $x_0 \in D$ is said to be a local maximal (resp. minimal) of a given function $y = f(x)$ if there exists a small neighborhood

$$\delta(x_0, D) = \{x \in D; |x - x_0| < \delta\}$$

such that

$$f(x_0) \geq f(x) \ (\text{resp. } f(x_0) \leq f(x))$$

for all $x \in \delta(x_0, D)$.

Examples

1. If $y = f(x)$ is monotonically increasing (resp. decreasing) over a subinterval $[a_1, x_0]$, but is monotonically decreasing (resp. increasing) over a subinterval $[x_0, b_1]$, then x_0 is, of course, a local maximal (resp. minimal).
2. -2 is a local minimal of $y = x^2 + 4x + 7$.
3. $\{2n\pi, n \in \mathbb{Z}\}$ are those local maximal points of $x = \cos t$ while $\{(2n + 1)\pi, n \in \mathbb{Z}\}$ are those local minimal points of $x = \cos t$.

4. $\left\{ \left(2n + \frac{1}{2}\right) \pi, n \in \mathbb{Z} \right\}$ are those local maximal points of $y = \sin t$ while $\left\{ \left(2n - \frac{1}{2}\right) \pi, n \in \mathbb{Z} \right\}$ are those local minimal points of $y = \sin t$.

1.2. *Continuity*

Intuitively and geometrically speaking, a function $y = f(x)$ defined over an interval $[a, b]$ is *continuous* if the graph of $y = f(x)$ is a continuous (i.e., unbroken) curve. From the viewpoint of functions, it roughly means that there are *no abrupt changes* of functional values, or to put it in positive tone, the change of functional values (which is measured by $|f(x_1) - f(x_2)|$) is small provided the change in x is *sufficiently* small. This suggests the following definition of the *localized concept of continuity* at a point.

Definition. A function $y = f(x)$ is said to be *continuous* at $x = a$ if, for any given $\varepsilon > 0$ (no matter how small), there always exists a (sufficiently small) $\delta > 0$ such that

$$|x - a| < \delta \Rightarrow |f(x) - f(a)| < \varepsilon .$$

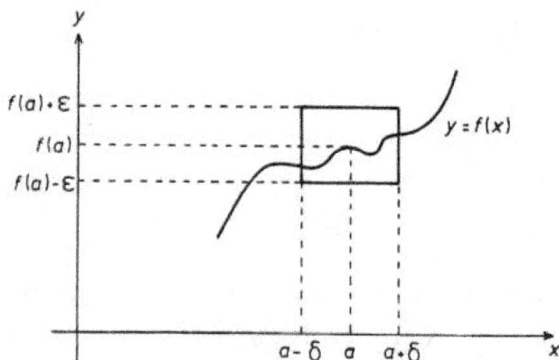

Fig. 17

Geometrically speaking, the above implication of inequalities simply means that a sufficiently small segment of the graph of $y = f(x)$

over $(a - \delta, a + \delta)$ lies completely inside the rectangular neighborhood of $(a, f(a))$ indicated in Fig. 17.

Definition. If a function $y = f(x)$ is continuous at every point of an interval, then it is said to be a continuous function over that interval.

Remark. Although the above definition of *continuity* is intuitively natural, the concept of continuity does involve some subtleties which are closely related to the concept of limit and the analytical formulation of the continuity of a straight line (cf. Chapter 3). Therefore, we shall wait until then for further clarification of such subtle points. Instead, we shall simply conclude this subsection by exhibiting some examples of *discontinuity*.

Example 1. Let $y = [x]$ be the largest integer $\leq x$ (called the integral part of x). It is a step function with discontinuity at every integer. The graph of $y = [x]$ looks like an ascending step of width one. It has a jump of 1 at each integral place.

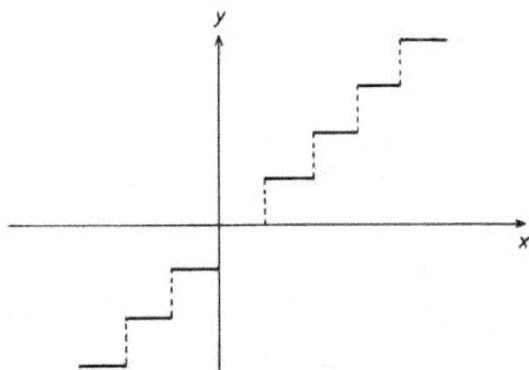

Fig. 18

Example 2. $y = \tan x = \frac{\sin x}{\cos x}$. The function $y = \tan x$ is, in fact, undefined at those points $x = \left(n + \frac{1}{2}\right)\pi$, $n \in \mathbb{Z}$. Hence, it is, of course, discontinuous at those points. Moreover, it is not difficult to see that the values of $y = \tan x$ becomes very large positive numbers as x approach $\left(n + \frac{1}{2}\right)\pi$ from the left-hand side while its values become negative numbers of very large absolute values as x approaches $\left(n + \frac{1}{2}\right)\pi$ from the right-hand side.

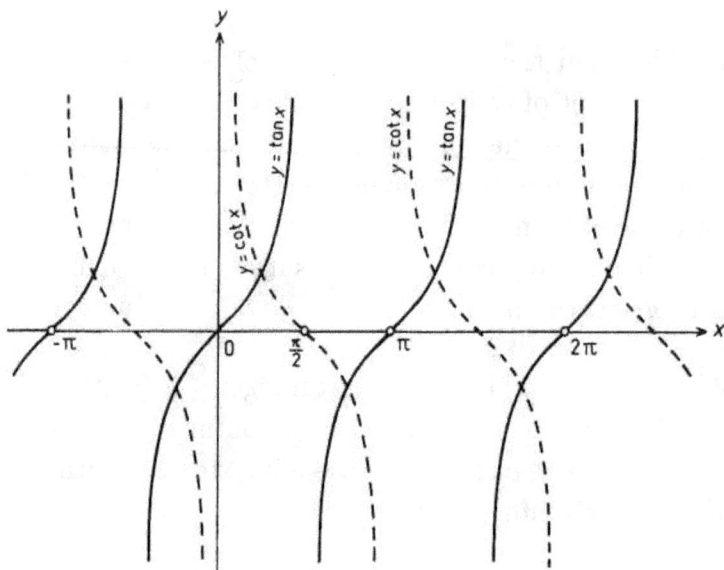

Fig. 19

Example 3. $y = \sin\frac{1}{x}$ (undefined at $x = 0$). In the neighborhood of $x = 0$, the functional values of $y = \sin\frac{1}{x}$ oscillate between $+1$ and -1 with rapidly quickening frequency. Hence it has a rather complicated type of discontinuity at $x = 0$.

Example 4. Define $y = f(x)$ by setting

$$f(x) = x\sin\frac{1}{x} \text{ if } x \neq 0 \text{ and } f(0) = 0. \tag{1}$$

Then it is an everywhere continuous function (including $x = 0$).

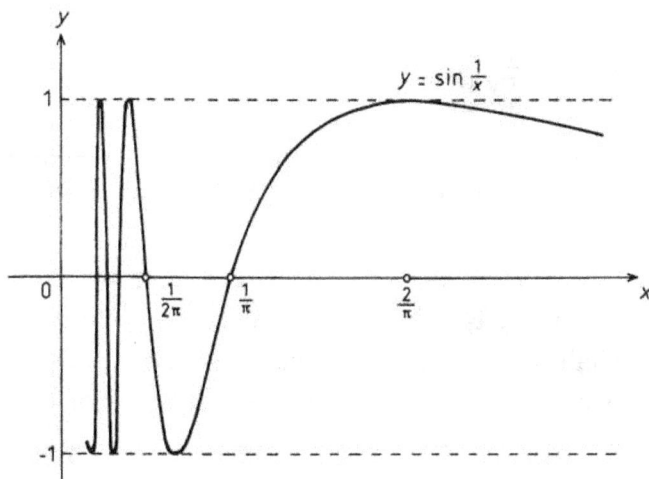

Fig. 20

Exercises

1. Verify the continuity of the above function of Example 4 at $x = 0$.
2. Roughly sketch the graph of the above function.
3. Show that $y = x^2$ is everywhere continuous.
4. Show that $y = \sin x$ is everywhere continuous.
5. Sketch the graph of $y = \sin x + \cos x$.
   [Hint: $\sin x + \cos x = \sqrt{2}\sin(x + \frac{\pi}{4})$.]
6. Sketch the graph of $y = \sin x \cdot \cos x$.
   [Hint: $\sin x \cos x = \frac{1}{2}\sin 2x$.]
7. Sketch the graph of $y = 3\sin x + 4\cos x$.
8. Sketch the graph of $y = x(x^2 - 1)$.
9. Find the extremal points and the extremal values of the following functions
   (i) $f(x) = x^2 + 2bx + c$.
   (ii) $f(x) = 3\sin x + 4\cos x$.
   (iii) $f(x) = 3\sin x - 4\cos x$.
10. Sketch the graph of

(i) $y = x \sin \frac{1}{x}$.

(ii) $y = x^2 \sin \frac{1}{x}$.

---

§ 2. Rate of Change and Sum of Total Effect of Functions

The *monotonicity* and the *continuity* that we discussed briefly in Sec. 1 are two basic qualitative features of functions. In this section, we shall proceed to study some basic *quantitative properties* of functions. Among various quantitative properties of functions, what are the most basic ones? At this junction, it is natural and also necessary to first look at some concrete examples.

Example 1 (Income tax). Every year the congress-persons take it as their solemn duty to define the income tax rate while each tax-payer is required to pay an amount called one's income tax.

Example 2 (Speed and mileage). Imagine that an impatient driver is speeding along a highway, not noticing that he or she has already been tailed by a police car. The policeman is checking the *speed* while the driver is mainly concerned about the mileage, e.g., how far has already been travelled and how much more remains toward the destination.

Example 3 (Interest and interest rate). In selecting the bank and the type of savings account, the primary concern is, of course, the interest rate.

Example 4. The electricity bill is based upon the total amount of electricity consumption of the household while the rate of electricity consumption is changing according to the number of watts in usage.

The above simple examples already demonstrate the basic importance of the following two quantitative properties of functions, namely

(i) the *rate of change* such as speed, interest rate and tax rate, etc.

(ii) the *sum of total effect* of a variable process such as the total mileage, the amount of electricity consumption, etc.

Indeed, the above two properties are exactly the most fundamental quantitative properties of functions and the study of these two properties naturally leads to the *differential calculus* and the *integral calculus* which constitutes the main framework of calculus. (Actually, the name of calculus also comes from the above fact.)

The precise analytical definition of the concept of the "*rate of change*" and the "*sum of total effect*" require the application of approximation procedure and the concept of limit (cf. Chapter 3). What we are going to discuss in this section is a thorough analysis of their intuitive contents, thus establishing the basic *comparison principles* which will provide the intuitive foundation of their precise analytical definitions, to be discussed in the next chapter.

2.1. *The rate of change of a piecewise linear function*

For a given linear function $y = mx + k$, if the value of the independent variable x is changed from x_1 to x_2, the corresponding value of the dependent variable y is changed from $y_1 = mx_1 + k$ to $y_2 = mx_2 + k$. Therefore, the ratio between their respective changes, namely

$$\frac{y_2 - y_1}{x_2 - x_1} = \frac{(mx_2 + k) - (mx_1 + k)}{x_2 - x_1} = \frac{m(x_2 - x_1)}{x_2 - x_1} = m \qquad (2)$$

is the constant m. Hence, the rate of change of the linear function $y = mx + k$ is clearly just the constant m, or rather, the constant function with value fixed at m.

Slightly more general, let $y = f(x)$ be a piecewise linear function, namely, there is a suitable subdivision of $[a, b]$, say

$$[a, b] = [a_0, a_1] \cup [a_1, a_2] \cup \cdots \cup [a_{i-1}, a_i] \cup \cdots \cup [a_{l-1}, a_l]$$

such that the restriction of $y = f(x)$ to each closed subinterval $[a_{i-1}, a_i]$, $1 \leq i \leq l$, is expressible by a suitable linear function. Therefore, the graph of $y = f(x)$ restricted to $[a_{i-1}, a_i]$ is a straight segment joining $(a_{i-1}, f(a_{i-1}))$ to $(a_i, f(a_i))$, whose slope is given by

$$m_i = \frac{f(a_i) - f(a_{i-1})}{a_i - a_{i-1}}. \tag{3}$$

Hence, the rate of change of the above piecewise linear function is equal to m_i for $x \in (a_{i-1}, a_i)$. Thus, the function which records the rate of change of the above piecewise linear function is exactly the step function (i.e., piecewise constant function) which takes the constant value m_i over the *open* interval (a_{i-1}, a_i).

2.2. *The sum of total effect of a piecewise constant process*

First of all, what should be the *sum of total effect* of a *constant* process is quite obvious and simple to calculate. For example, if one is driving at a constant speed v_0 then the total mileage one travels over the time interval $a \leq t \leq b$ is simply equal to $v_0(b - a)$. Similarly, if the rate of electricity consumption of a household was kept at a constant of w_0 watts over a time period $[t_1, t_2]$, then the total amount of electricity consumption over this period is also given by $w_0(t_2 - t_1)$.

In a slightly more general situation, suppose the whole period of a month can be suitably subdivided into l subintervals of time such that the rate of electricity consumption was kept at a constant w_i for the i-th subinterval (t_{i-1}, t_i). Then, the total amount of electricity consumption of the whole month is, of course, equal to the sum of that of each subinterval of time, namely

$$E = \sum_{i=1}^{l} E_i = \sum_{i=1}^{l} w_i(t_i - t_{i-1}).$$

Formulating in terms of functions, the sum of total effect of a piecewise constant process, such as the above example of electricity consumption, can always be computed by the above formula. Adopting the terminology of calculus, we shall call the above sum the definite integral of the step function $f(t)$ over the interval $[a, b]$ and denote it by the regular general notation of integration, namely $\int_a^b f(t)dt$.

Definition. Let $y = f(x)$ be a step function with constant value c_i on the i-th subinterval (a_{i-1}, a_i). Then the *definite integral* of $f(x)$ over the interval (a_0, a_l) is defined to be the following sum, namely

$$\int_{a_0}^{a_l} f(x)dx = \sum_{i=1}^{l} c_i(a_i - a_{i-1}). \tag{4}$$

For example, if the speed (resp. the rate of electricity consumption) is such a step function, then the above definite integral is exactly the total mileage (resp. total amount of electricity consumption) over the time interval (a_0, a_l). Geometrically, one may interpret each summand $c_i(a_i - a_{i-1})$ as the *oriented area* (i.e., negative if it is below the x-axis) of the rectangle between the i-th step and the x-axis (cf. Fig. 21). Therefore, the definite integral can be interpreted as the total area bounded by the graph of $y = f(x)$, the x-axis and the two vertical lines of $x = a_0$ and $x = a_l$.

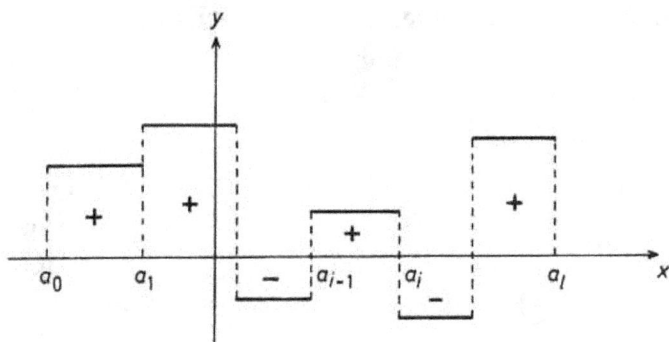

Fig. 21

Remark. The family of *piecewise linear* (resp. *step*) functions constitutes a *basic* family of functions whose "*rates of change*" (resp.

sums of total effect, or *definite integrals*) are easy to see and simple to compute. However, they constitute the basis for applying the *methodology of approximation* to provide an analytical definition of the rate of change (resp. definite integral) for functions in general, just as the way Eudoxus did more than two millenniums ago in the understanding of ratios between non-commensurable pairs of intervals.

The first step of such an approach is to analyze the intuitive content of the rate of change (resp. definite integral) in order to establish a correct comparison principle on the rate of change (resp. definite integral).

2.3. Comparison principles of the rate of change and the definite integral of functions

Analysis of the intuitive content of the rate of change

1. The rate of change of a given function $y = f(x)$ at a given point $x = x_0$ is a *local property* which *only* depends on the functional values in an arbitrarily small neighborhood of x_0, i.e., $(x_0 - \delta, x_0 + \delta)$ with δ an arbitrary small positive number. Moreover, if $y = f(x)$ is monotonically increasing (resp. decreasing) in a vicinity of x_0, then the rate of change of $y = f(x)$ at x_0 must be non-negative (resp. non-positive). For example, the rate of change of $y = x^2$ must be non-positive for $x_0 < 0$ but non-negative for $x_0 > 0$ while the rate of change of $y = x^3$ must be non-negative for all x_0.

2. Let $f_1(x)$ and $f_2(x)$ be two given functions and $g(x)$ be the difference function, namely, $g(x) = f_1(x) - f_2(x)$. For example, let x be the time and $f_1(x)$, $f_2(x)$ record the distance travelled by cars No. 1 and No. 2 respectively. Then $g(x)$ records the relative distance between them. The rate of change of the relative distance is the relative speed which is exactly the difference between the speeds of cars No. 1 and No. 2. Therefore, the rate of change of the *difference function* of $f_1(x)$ and $f_2(x)$ is always equal to the difference of the rates of change of $f_1(x)$ and $f_2(x)$.

The above analysis naturally leads to the following comparison principle of rate of change.

Comparison principle of rate of change. Let $f_1(x)$ and $f_2(x)$ be two continuous functions with $f_1(x_0) = f_2(x_0)$. If there exists a sufficiently small neighborhood of x_0, say $(x_0 - \delta, x_0 + \delta)$, such that

$$\begin{cases} f_1(x) - f_2(x) < 0 \text{ (resp. } > 0) \text{ for } x_0 - \delta < x < x_0 \\ f_1(x) - f_2(x) > 0 \text{ (resp. } < 0) \text{ for } x_0 < x < x + \delta \end{cases}, \quad (5)$$

then the rate of change of $f_1(x)$ at $x = x_0$ must be at least (resp. at most) equal to that of $f_2(x)$.

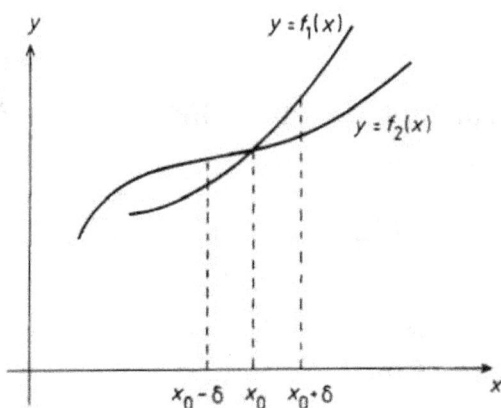

Fig. 22

Remark. Notice that, as far as our discussion is concerned, the concept of the rate of change of a general function (i.e., other than those piecewise linear ones) is something *yet to be defined*. However, the above *comparison* principle is *compelled by the intuitive meaning of the rate of change!*

Next let us analyze the intuitive content of the definite integral, namely, the sum of total effect.

Analysis

1. The definite integral is, of course, a *global* property which depends on the functional values of the given function $y = f(x)$ over the whole interval $[a, b]$.

2. Suppose that the interval $[a, b]$ is subdivided into the union of l subintervals by

$$a = a_0 < a_1 < a_2 < \cdots < a_{i-1} < a_i < \cdots < a_l = b \,,$$

then the intuitive meaning of definite integral compels that

$$\int_a^b f(x)dx = \sum_{i=1}^{l} \int_{a_{i-1}}^{a_i} f(x)dx. \tag{6}$$

Comparison principle of definite integral. Let $f(x)$ and $g(x)$ be two functions defined on $[a, b]$. If $f(x) \geq g(x)$ for all $a \leq x \leq b$, then

$$\int_a^b f(x)dx \geq \int_a^b g(x)dx. \tag{7}$$

(The above comparison principle is clearly compelled by the intuitive meaning of the sum of total effect.)

2.4. *Some preliminary applications of the above comparison principles*

As a prelude to the systematic application of approximation methodology to the study of the rate of change and the definite integral, we shall conclude this section by some preliminary applications of the above comparison principles to the study of the rate of change as well as the definite integral of such simple polynomial functions as $y = x^2$, $y = x^3$, etc.

Example 1. What should be the rate of change of the function $y = x^2$ at $x = a$?

Solution. Notice that the rate of change is already clearly defined for linear functions, i.e., polynomial functions of degree ≤ 1, while the rate of change of $y = x^2$ is something *yet to be defined*. Therefore, one naturally tries to apply the comparison principle (of rate of change) to $y = x^2$ with those of linear functions. Let us compare the difference of functional values between $y = x^2$ and $y = a^2 + m(x - a)$ over a sufficiently small neighborhood of a, say $(a - \delta, a + \delta)$, namely

$$
\begin{aligned}
x^2 - [a^2 + m(x - a)] &= (x - a)[(x + a) - m] \\
&= (x - a)[(2a - m) + (x - a)].
\end{aligned}
\tag{8}
$$

If $(2a - m) > 0$ (resp. < 0) and $|x - a| < |2a - m|$, then

$$
\begin{cases}
x^2 - [a^2 + m(x - a)] > 0 \text{ (resp. } < 0) \text{ for } x - a > 0 \\
x^2 - [a^2 + m(x - a)] < 0 \text{ (resp. } > 0) \text{ for } x - a < 0.
\end{cases}
\tag{9}
$$

Hence, by the comparison principle, the rate of change of $y = x^2$ at $x = a$ must be $\geq m$ (resp. $\leq m$) for any $m < 2a$ (resp. $m > 2a$). Thus, the only possibility is that the rate of change of $y = x^2$ at $x = a$ is equal to $2a$! In other words, to define the rate of change of $y = x^2$ at $x = a$ to be equal to $2a$ is the *only reasonable choice*, if it is, at all possible, to be defined.

Example 2. What should be the rate of change of $y = x^3$ at $x = a$?

Solution. Set $x = a + t$ and $f_1(t) = (a + t)^3$. Then the rate of change of $y = x^3$ at $x = a$ is just the rate of change of $f_1(t)$ at $t = 0$. Apply the comparison principle to $f_1(t)$ and $f_2(t) = a^3 + mt$ at $t = 0$, namely

$$
\begin{aligned}
f_1(t) - f_2(t) &= (a + t)^3 - (a^3 + mt) \\
&= (3a^2 - m)t + 3at^2 + t^3 \\
&= t \cdot [(3a^2 - m) + 3at + t^2].
\end{aligned}
\tag{10}
$$

Therefore, for a sufficiently small $|t|$ and $(3a^2 - m) \neq 0$

$$
(3a^2 - m) + 3at + t^2
\tag{11}
$$

always has the same sign as that of $(3a^2 - m)$. Hence

$$3a^2 - m \begin{cases} > 0 \\ < 0 \end{cases} \Rightarrow f_1(t) - f_2(t) \begin{cases} \text{has the same sign of } t \\ \text{has the opposite sign of } t. \end{cases} \qquad (12)$$

Thus the rate of change of $f_1(t)$ at $t = 0$ must be $\geq m$ (resp. $\leq m$) for any $m < 3a^2$ (resp. $m > 3a^2$), and hence the only possibility is that the rate of change of $f_1(t)$ at $t = 0$ is equal to $3a^2$.

Remark. Similar computation and application of the comparison principle will show that the rate of change of $f_1(t) = (a + t)^n = a^n + na^{n-1}t + \frac{n(n-1)}{2!}a^{n-2}t^2 + \cdots + t^n$ at $t = 0$ is equal to na^{n-1}.

Example 3. $\int_0^b x\,dx = ?$

Solution. Geometrically, $\int_0^b x\,dx$ should be equal to the area of the right-angled triangle $\triangle OBA$ indicated in Fig. 23 which is equal to $\frac{1}{2}b^2$. However, we shall use the comparison principle of definite integral to show that $\int_0^b x\,dx = \frac{1}{2}b^2$.

Let us first subdivide the interval $[0, b]$ into n subintervals of equal length. As indicated in Fig. 23 let $G_n(x)$ and $g_n(x)$ be the step functions defined by setting

$$\begin{cases} G_n(x) = \frac{i}{n}b & \text{for } \frac{i-1}{n}b < x \leq \frac{i}{n}b, \\ g_n(x) = \frac{i-1}{n}b & \text{for } \frac{i-1}{n}b \leq x < \frac{i}{n}b \end{cases} \quad (1 \leq i \leq n). \qquad (13)$$

Then, one clearly has

$$g_n(x) \leq x \leq G_n(x) \text{ for all } 0 \leq x \leq b. \qquad (14)$$

Hence, the yet to be determined number $\int_0^b x\,dx$ should satisfy the following inequalities by the comparison principle, namely

$$\int_0^b g_n(x)\,dx \leq \int_0^b x\,dx \leq \int_0^b G_n(x)\,dx. \qquad (15)$$

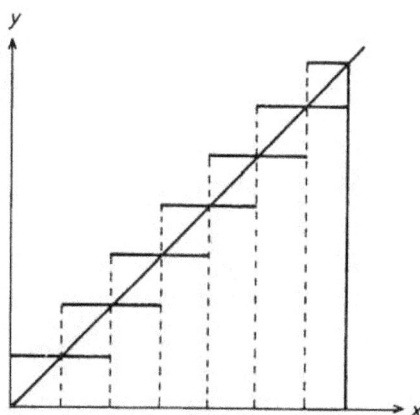

Fig. 23

Straightforward computation shows that

$$\int_0^b G_n(x)dx = \sum_{i=1}^n \frac{i}{n}b \cdot \frac{b}{n} = \frac{b^2}{n^2}\sum_{i=1}^n i$$

$$= \frac{b^2}{n^2}\frac{1}{2}n(n+1) = \frac{b^2}{2}\left(1 + \frac{1}{n}\right)$$

$$\int_0^b g_n(x)dx = \sum_{i=1}^n \frac{i-1}{n}b\frac{b}{n} = \frac{b^2}{n^2}\sum_{i=1}^n (i-1)$$

$$= \frac{b^2}{n^2}\frac{1}{2}n(n-1) = \frac{b^2}{2}\left(1 - \frac{1}{n}\right). \tag{16}$$

Thus

$$\frac{b^2}{2}\left(1 - \frac{1}{n}\right) \le \int_0^b x\,dx \le \frac{b^2}{2}\left(1 + \frac{1}{n}\right) \tag{17}$$

holds for any positive integer n, no matter how large it may be. Hence $\frac{b^2}{2}$ is the only possible value for $\int_0^b x\,dx$ that satisfies the above inequalities compelled by the comparison principle! This proves that

$$\int_0^b x\,dx = \frac{b^2}{2}. \tag{18}$$

Example 4. $\int_0^b x^2 dx = ?$

Solution. Again, first equally subdivide the interval $[0, b]$ into n subintervals of length $\frac{b}{n}$ and then set $G_n(x)$ and $g_n(x)$ to be the following step functions, namely

$$G_n(x) = \left(\frac{i}{n}b\right)^2 = \frac{b^2}{n^2}i^2 \text{ for } \frac{i-1}{n}b < x \le \frac{i}{n}b,$$
$$g_n(x) = \left(\frac{i-1}{n}b\right)^2 = \frac{b^2}{n^2}(i-1)^2 \text{ for } \frac{i-1}{n}b \le x < \frac{i}{n}b \quad (1 \le i \le n). \quad (19)$$

Then, again, one has

$$g_n(x) \le x^2 \le G_n(x) \text{ for all } a \le x \le b \tag{20}$$

and hence, by the comparison principle

$$
\begin{array}{ccc}
\int_0^b g_n(x)dx & \le \int_0^b x^2 dx \le & \int_0^b G_n(x)dx \\
\| & & \| \\
\sum_{i=1}^n \frac{b^3}{n^3}(i-1)^2 & & \sum_{i=1}^n \frac{b^3}{n^3}i^2.
\end{array}
\tag{21}
$$

Moreover, it is not difficult to verify by the mathematical induction the following summation formula

$$\sum_{i=1}^n i^2 = \frac{1}{6}n(n+1)(2n+1). \tag{22}$$

Therefore

$$\frac{b^3}{3}\left(1 - \frac{1}{n}\right)\left(1 - \frac{1}{2n}\right) \le \int_0^b x^2 dx \le \frac{b^3}{3}\left(1 + \frac{1}{n}\right)\left(1 + \frac{1}{2n}\right) \tag{23}$$

holds for all $n \in \mathbb{N}$. Since $\frac{b^3}{3}$ is the only possible number satisfies the above lower and upper bounds for all n, the above inequalities compel that

$$\int_0^b x^2 dx = \frac{b^3}{3}. \tag{24}$$

Remark. It is not difficult to use similar computation and the comparison principle to prove that

$$\int_0^b x^k dx = \frac{1}{(k+1)} b^{k+1}. \tag{25}$$

Exercises

1. $\int_a^b x\, dx =$?
2. $\int_a^b x^2 dx =$?
3. $\int_a^b x^3 dx =$?
4. $\int_a^b (x + x^2 - x^3) dx =$?
5. What is the rate of change of $y = c_0 + c_1 x + c_2 x^2$ at $x = a$?
6. What is the rate of change of $y = -x^3 + x^2$ at $x = -1$?
7. Let $y = f(x)$ be a step function (i.e., piecewise constant function). Set

   $$G(t) = \int_{a_0}^t f(x) dx$$

   and let t vary from a_0 to $a_l = b$. What can one say about the function $G(t)$?
8. What is the rate of change of $y = x^3 + bx^2 + cx + d$ at an arbitrarily given point $x = x_0$?
9. Using the result obtained in Exercise 8 to determine the algebraic condition on b and c which makes $y = x^3 + bx^2 + cx + d$ have local extremals.

   [Hint: The rate of change of the above cubic polynomial at x_0 will be a quadratic polynomial in x_0. What is the condition on its coefficients that a quadratic polynomial has distinct real roots?]
10. Determine the local maximal and/or local minimal of $y = x^3 + bx^2 + cx + d$ satisfying the condition obtained in Exercise 9.

CHAPTER 3

Approximation and Limit

Approximation is the fundamental methodology which plays an important role throughout the entire realm of calculus while *limit* is the basic concept that naturally arises from applications of approximation methodology.

The basic idea of approximation is to use a family of simpler or better-known objects to approximate a certain more general or new type of objects so that the *"difference"* on certain pertinent quantitative aspects can be made to be as small as one wishes. Generally speaking, this is often an effective methodology which enables us to reduce the study of the latter to that of the former. Historically, such an ingenious methodology of approximation was first invented by the great geometer Eudoxus to overcome the difficulty posed by the existence of *non-commensurable quantities* to the entire foundation of geometry. His basic idea was, in essence, to approximate irrational numbers by the rational numbers!

§ 1. Approximation and Limit of Sequences

1.1. *Eudoxian principle and approximation methodology*

Recall that the *real number system* $\mathbb{R}$ is the mathematical system created for the purpose of studying quantities of measurement type such as length, area, angle weight etc. The ratios between commensurable pairs are rational numbers which are conceptually rather simple and computationally easily reducible to that of integers. However, the ratios between *non-commensurable pairs* are *irrational numbers* which are conceptually more sophisticated and technically much more difficult to handle. For example, the existence of irrational ratios had baffled the geometers of antiquity for more than half a century, namely, after the discovery of non-commensurable pairs of intervals by Hippasus up until the invention of approximation methodology by Eudoxus. Anyhow, approximation and limit are from the very beginning an intrinsic part of the real number system and the real number system is also the canonical place to begin our discussion of limit concept and approximation methodology.

As it has already been briefly discussed in Chapter 1, the enunciation of Eudoxian principle and its preliminary application to *rebuild* the "shattered foundation of quantitative geometry" was a resounding success, a great beginning of the approximation methodology which has far-reaching applications throughout calculus and beyond! The underlying idea of Eudoxian principle is to approximate irrationals by rationals. Let us take the irrational $\sqrt{2}$ as an example. Eudoxian principle enunciates that $\sqrt{2}$ is the *unique* real number which is larger than any positive rational, $\frac{m}{n}$, whose square is less than 2, and is smaller than any positive rational, $\frac{m}{n}$, whose square is more than 2. Such a characterization of $\sqrt{2}$ is clearly correct but it certainly looks rather cumbersome. From the approximation viewpoint, the above characterization of $\sqrt{2}$ can be reformulated as follows:

Let $\{a_n, \ n \in \mathbb{N}\}$ and $\{b_n, \ n \in \mathbb{N}\}$ be two sequences of positive rationals such that

$$a_1 \leq a_2 \leq \cdots \leq a_n \leq a_{n+1} \leq \cdots < \sqrt{2} < \cdots$$
$$\leq b_{n+1} \leq b_n \leq \cdots \leq b_2 \leq b_1 \tag{1}$$

and moreover $b_n - a_n$ becomes as small as one wants provided n is sufficiently large. Then $\sqrt{2}$ is the unique number that lies between all pairs of a_n and b_n, namely

$$a_n < x < b_n \text{ for all } n \Leftrightarrow x = \sqrt{2}. \tag{2}$$

We shall call such a sequence $\{a_n, \ n \in \mathbf{N}\}$ (resp. $\{b_n, \ n \in \mathbf{N}\}$) an increasing (resp. decreasing) approximating sequence of $\sqrt{2}$; a_n (resp. b_n) is the n-th approximate value from below (resp. above) while their errors, i.e., $\sqrt{2} - a_n$ and $\sqrt{2} - b_n$ are clearly smaller than $b_n - a_n$.

This suggests the following terminology of sequences which will facilitate the discussion and the applications of approximation and limit.

1.2. Sequences

Definition. A sequence of real numbers assigns a real number to each natural number $n \in \mathbf{N}$ and calls it the n-th term of the given sequence. (In most cases, it is actually the n-th approximate value.)

Usually, we use $\{a_n, \ n \in \mathbf{N}\}$ or simply $\{a_n\}$ to denote the sequence whose n-th term is a_n. The following are some simple examples of sequences.

Examples of sequences 1. $\{1, 2, 3, \ldots, n, \ldots\}$ is a sequence whose n-th term is n. 2. $\{1, -1, 1, -1, \ldots, 1, -1, \ldots\}$ is a sequence whose n-th term is equal to $(-1)^{n+1}$.

3. $\left\{1, \frac{1}{2}, \frac{1}{3}, \ldots, \frac{1}{n}, \ldots\right\}$ is a sequence whose n-th term is equal to $\frac{1}{n}$.

4. $\{a_n = \text{ the maximal one among all numbers of } n\text{-digits } < \sqrt{2}\}$, e.g. $a_1 = 1.4$, $a_2 = 1.41$, $a_3 = 1.414$, $a_4 = 1.4142$, $a_5 = 1.41421$, etc.

5. $\{b_n = \text{ the minimal one among all numbers of } n\text{-digits } > \sqrt{2}\}$, e.g. $b_1 = 1.5$, $b_2 = 1.42$, $b_3 = 1.415$, $b_4 = 1.4143$, $b_5 = 1.41422\ldots$, and in general $b_n = a_n + \left(\frac{1}{10}\right)^n$.

6. $\{c, c, c, \ldots, c, \ldots\}$ is a constant sequence whose n-th term is always equal to c.

Generally speaking, a sequence, $\{a_n\}$, naturally arises in an approximation procedure whose n-th term is the n-th approximate value. For example, the sequence $\{a_n\}$ of Example 4 arises in the approximation of $\sqrt{2}$ by digitals from the left side and the sequence $\{b_n\}$ of Example 5 arises in the approximation of $\sqrt{2}$ by digitals from the right side.

Definition. A sequence $\{a_n\}$ is said to be an approximation sequence of α if the error, $|a_n - \alpha|$, can be made *as small as one wants* provided that n is *sufficiently large*. (In terms of formal mathematical language, this definition is usually rephrased as follows: To any given $\varepsilon > 0$, there always exists an integer N such that $|a_n - \alpha| < \varepsilon$ for all $n \geq N$.)

The above formal logical statement makes it clear that the desired smallness, i.e., $\varepsilon > 0$, must be given first; then one can determine how large the indices n is, in fact, sufficient.

Notation. We shall denote the above relationship between an approximation sequence, $\{a_n\}$, and the target of approximation, α, by $a_n \to \alpha$.

Examples

    1. If $\left\{a_n = \frac{1}{n}\right\}$, then $a_n \to 0$.

    2. If $\left\{a_n = \left(\frac{1}{10}\right)^n\right\}$, then $a_n \to 0$.

    3. If $\left\{a_n = \frac{K}{2^n}\right\}$, then $a_n \to 0$.

    4. Let $\{a_n\}$ and $\{b_n\}$ be respectively the previous Example 4 and Example 5. Then $a_n \to \sqrt{2}$ and $b_n \to \sqrt{2}$.

    5. $a_n \to \alpha$ if and only if $(a_n - \alpha) \to 0$.

    6. Suppose $\{a_n\}$ and $\{b_n\}$ are two sequences satisfying the following conditions, namely,

$$\begin{cases} a_1 \leq a_2 \leq \cdots \leq a_n \leq a_{n+1} \leq \cdots \leq \alpha \leq \\ \quad \cdots \leq b_{n+1} \leq b_n \leq \cdots \leq b_2 \leq b_1, \\ \text{and } (b_n - a_n) \to 0. \end{cases} \quad (3)$$

Then $a_n \to \alpha$ and $b_n \to \alpha$.

Proof. To any given $\varepsilon > 0$, it follows from the condition $(b_n - a_n) \to 0$

that there exists an integer N, such that

$$(b_n - a_n) = |(b_n - a_n) - 0| < \varepsilon \text{ for all } n \geq N. \tag{4}$$

Since $a_n \leq \alpha \leq b_n$, it is clear that

$$\left. \begin{array}{c} |a_n - \alpha| \\ |b_n - \alpha| \end{array} \right\} \leq (b_n - a_n) < \varepsilon \text{ for all } n \geq N. \tag{5}$$

This proves that $a_n \to \alpha$ and $b_n \to \alpha$. (We shall call such a pair of sequences $\{a_n\}, \{b_n\}$ as a pair of *approaching sequences* and α is called their *common limit*.)

In terms of the above definitions and terminology, it is convenient to reformulate the Eudoxian principle as follows:

Suppose that there exist $\{a_n\}, \{b_n\}$ and α, α' satisfying the following conditions:

$$\left\{ \begin{array}{l} a_1 \leq a_2 \leq \cdots \leq a_n \leq a_{n+1} \leq \cdots \leq \left\{ \begin{array}{c} \alpha \\ \alpha' \end{array} \right\} \leq \\ \qquad \cdots \leq b_{n+1} \leq b_n \leq \cdots \leq b_2 \leq b_1 \\ \text{and } (b_n - a_n) \to 0. \end{array} \right. \tag{6}$$

Then $\alpha = \alpha'$ because $|\alpha - \alpha'| < (b_n - a_n)$ for all n.

The above reformulation makes it clear that the Eudoxian principle is precisely the *uniqueness* aspect of the common limit of a pair of approaching sequences. Naturally, one should also investigate the *existence* aspect of the common limit of a pair of approaching sequences, namely,

Problem of Existence. Suppose $\{a_n\}$ and $\{b_n\}$ are a pair of approaching sequences, i.e.,

$$\left\{ \begin{array}{l} a_1 \leq a_2 \leq \cdots \leq a_n \leq a_{n+1} \\ \qquad \leq \cdots \leq b_{n+1} \leq b_n \leq \cdots \leq b_2 \leq b_1 \\ \text{and } (b_n - a_n) \to 0. \end{array} \right. \tag{7}$$

Is it true that there always exists a suitable real number α which is exactly the common limit of them, i.e., $a_n \leq \alpha \leq b_n$ for all n?

In fact, it is a *fundamental property of the real number system* that such a *common limit α always exists*. From the geometric interpretation of real numbers as the lengths of intervals, the above existence statement can actually be considered as the *analytical description* (or *formulation*) of the geometric intuition of the *continuity of a straight line*. (Geometrically, a "non-existence situation" indicates the occurrence of a "gap", while the continuity of a straight line means there is no gap at all!) We restate it as follows:

Continuity (or completeness) of the real number system. To any pair of approaching sequences $\{a_n\}$ and $\{b_n\}$, namely,

$$\begin{cases} a_1 \leq a_2 \leq \cdots \leq a_n \leq a_{n+1} \\ \qquad \leq \cdots \leq b_{n+1} \leq b_n \leq \cdots \leq b_2 \leq b_1 \\ \text{and } (b_n - a_n) \rightarrow 0, \end{cases} \qquad (8)$$

there always exists a real number α such that

$$a_n \leq \alpha \leq b_n \text{ for all } n.$$

Remark. The above property is *the fundamental existence statement* in the whole theory of mathematical analysis.

In fact, the proofs of all those basic existence theorems in calculus and beyond are all based upon it!

1.3. *Limit and an existence theorem*

Definition. α is said to be equal to the *limit* of a sequence $\{a_n\}$ (as $n \rightarrow \infty$) if and only if $\{a_n\}$ is an approximation sequence of α.

Remark. Technically, $\lim_{n\rightarrow\infty} a_n = \alpha$ and $a_n \rightarrow \alpha$ are simply two slightly different ways of paraphrasing the same relationship between the sequence $\{a_n\}$ and the number α. However, there are important differences both in viewpoint and in emphasis. When α is given first and one proceeds to construct a sequence $\{a_n\}$ to approach α, this is the viewpoint of approximation and the emphasis is on the *uniqueness* aspect. On the other hand, when a sequence $\{a_n\}$ is given first and one tries to determine whether there exists a suitable number α with $\lim_{n\to\infty} a_n = \alpha$ (i.e., to be approximated by a_n); this is the viewpoint of limit and the emphasis is mainly on the *existence* aspect. The following is a useful example of existence theorems on the limit of sequences. First, we need a few simple definitions.

Definition. A sequence $\{s_n\}$ is said to be (monotonically) *increasing* (resp. *decreasing*) if $s_n \leq s_{n+1}$ (resp. $s_n \geq s_{n+1}$) for all $n \in \mathbb{N}$.

Definition. A sequence $\{s_n\}$ is said to be bounded from above (resp. below) if there exists a constant K such that

$$s_n \leq K \text{ (resp. } s_n \geq K) \text{ for all } n \in \mathbb{N}. \tag{9}$$

Such a constant K is called an upper (resp. lower) bound of the sequence $\{s_n\}$.

Theorem 3.1. *If $\{s_n\}$ is an increasing (resp. decreasing) sequence and is bounded from above (resp. below), then $\lim_{n\to\infty} s_n$ exists.*

Proof. The proofs of the two cases are the same. Hence, we shall only prove the case that $\{s_n\}$ is increasing and bounded from above. Intuitively, the limit of such an increasing sequence $\{s_n\}$ should be the smallest upper bound of $\{s_n\}$. Technically, we shall need the fundamental existential statement of Sec. 1.2, namely, the continuity of the real number system $\mathbb{R}$, to establish the existence of $\lim_{n\to\infty} s_n$. Therefore, the basic idea of our proof is to construct a pair of approaching sequences $\{a_n\}$ and $\{b_n\}$ which will have the yet-to-be found limit value α as their common limit. The following is an inductive procedure of constructing such a pair of approaching sequences by the method of *successive bisection*.

To begin with, set $a_1 = s_1$ and $b_1 = K$ (an arbitrary upper bound of $\{s_n\}$). We bisect the interval $[a_1, b_1]$ and select a half to be the next interval $[a_2, b_2]$ according to the following rule, namely,

$$\begin{cases} \text{if } \tfrac{1}{2}(a_1 + b_1) \text{ is still an upper bound} \\ \qquad\qquad \text{then the first half is } [a_2, b_2], \\ \text{if } \tfrac{1}{2}(a_1 + b_1) \text{ is not an upper bound} \\ \qquad\qquad \text{then the seccond half is } [a_2, b_2]. \end{cases}$$

Observe that the above way of selection of the half interval, $[a_2, b_2]$, preserves the following two crucial properties, namely,

(i) b_2 is again an upper bound and (ii) $[a_2, b_2] \cap \{s_n\} \neq \emptyset$.

Inductively, suppose one already made the selection of the m-th subinterval $[a_m, b_m]$ preserving the above two properties. Then, by the same way of selection will be obtained a half interval, i.e., the $(m+1)$-th subinterval, $[a_{m+1}, b_{m+1}]$, which, again, inherits the above two properties. By the above construction, it is clear that $\{a_m\}$ is an increasing sequence, $\{b_m\}$ is a decreasing sequence, and $(b_m - a_m) = \frac{1}{2^{m-1}}(b_1 - a_1) \to 0$. Moreover, for all $m \in \mathbb{N}$,

(i) b_m is an upper bound of $\{s_n\}$ and (ii) $[a_m, b_m] \cap \{s_n\} \neq \emptyset$.

Let α be the common limit of the above pair of approaching sequences, i.e., $a_m \to \alpha$ and $b_m \to \alpha$. We shall verify that $s_n \to \alpha$.

To any given $\varepsilon > 0$, it follows from $a_m \to \alpha$ and $b_m \to \alpha$ that there exists a m_0 such that

$$|a_m - \alpha| < \varepsilon \text{ and } |b_m - \alpha| < \varepsilon \text{ for all } m \geq m_0, \qquad (10)$$

in particular, $[a_{m_0}, b_{m_0}] \subset (\alpha - \varepsilon, \alpha + \varepsilon)$. On the other hand, there exists an integer N such that $s_N \in [a_{m_0}, b_{m_0}]$ (by construction,

$[a_{m_0}, b_{m_0}] \cap \{s_n\} \neq \emptyset$). Observe that b_{m_0} is an upper bound and $\{s_n\}$ is increasing. Hence, for all $n \geq N$,

$$\alpha - \varepsilon \leq a_{m_0} \leq s_N \leq s_n \leq b_{m_0} < \alpha + \varepsilon, \tag{11}$$

namely,

$$|s_n - \alpha| < \varepsilon. \tag{12}$$

This proves that α is the limit of $\{s_n\}$, i.e., $\alpha = \lim_{n \to \infty} s_n$. $\qquad\square$

Remarks

(i) For a reader at his or her first reading, the above proof may be a bit puzzling or even create some uneasiness.

In fact, this is quite normal for such a "first encounter". After all, it is a new type of proof dealing with some new concepts and new phenomena. For example, the concepts of limit and of the continuity of real number systems are certainly "new acquaintances" and the construction of the pair of approaching sequences $\{a_n\}$ and $\{b_n\}$ is a new phenomenon actually involving infinitely many inductive steps.

(ii) In analyzing the above proof, it is instructive to distinguish the special feature from the general feature. Suppose that α is the place that $\lim_{n \to \infty} s_n$ should be situated and $[\alpha - \varepsilon, \alpha + \varepsilon]$ is an arbitrarily small neighborhood of α. Then it follows from the assumption of increasing monotonicity of $\{s_n\}$ that $\alpha + \varepsilon$ must be an upper bound of $\{s_n\}$ and $[\alpha - \varepsilon, \alpha + \varepsilon] \cap \{s_n\} \neq \emptyset$. This is exactly the special, characteristic properties of an arbitrarily small neighborhood of the limit place of such a sequence which plays the role of selection criterion. On the other hand, the basic idea of "zero-in" by the nest of intervals $\{[a_m, b_m], m \in \mathbf{N}\}$ onto the yet-to-be-proved limit place is perfectly general. It is a general technique fitting for obtaining the existence of a point with certain desired properties by reducing it to the fundamental existence statement, namely, the continuity of the real line.

(iii) Intuitively, the above proof amounts to establish the existence of the limit place by approximation. The m-th subinterval

$[a_m, b_m]$ can be regarded as the m-th stage of the approximation. For the specific problem of Theorem 3.1, the above special property which characterizes an arbitrary small neighborhood of the target-place naturally provides the much needed guiding criterion for the inductive selection of subintervals, $\{[a_m, b_m], \ m \in \mathbb{N}\}$.

Exercises

To verify the existence of limit for the following sequences:

1. $a_1 = \sqrt{2}, a_2 = \sqrt{2 + \sqrt{2}}, a_3 = \sqrt{2 + \sqrt{2 + \sqrt{2}}}, \ldots$ and inductively $a_{n+1} = \sqrt{2 + a_n}$.

2. $a_1 = \sqrt{2}, a_2 = \sqrt{2 \cdot \sqrt{2}}, a_3 = \sqrt{2 \cdot \sqrt{2 \cdot \sqrt{2}}}, \ldots$ and inductively $a_{n+1} = \sqrt{2 \cdot a_n}$.

3. $\left\{ a_n = \frac{1}{n+1} + \frac{1}{n+2} + \cdots + \frac{1}{2n}, \ n = 1, 2, 3, \ldots \right\}$.

4. $\left\{ a_n = \left(1 + \frac{1}{n}\right)^n, \ n = 1, 2, 3, \ldots \right\}$.

5. $\left\{ b_n = \left(1 + \frac{1}{n}\right)^{n+1}, \ n = 1, 2, 3, \ldots \right\}$.

   [Hint: Compute $\frac{a_{n+1}}{a_n}$ and $\frac{b_n}{b_{n+1}}$ for Exercises 4 and 5 and make use of the inequality $\left(1 - \frac{1}{(n+1)^2}\right)^{n+1} > 1 - \frac{1}{n+1}$ to show that a_n (resp. b_n) are monotonically increasing (resp. decreasing).]

6. Suppose that $a_1 < b_1$ are two given positive real numbers. Set

$$a_2 = \sqrt{a_1 b_1}, \quad b_2 = \frac{1}{2}(a_1 + b_1)$$

$$a_3 = \sqrt{a_2 b_2}, \quad b_3 = \frac{1}{2}(a_2 + b_2)$$

and inductively

$$a_{n+1} = \sqrt{a_n b_n}, \quad b_{n+1} = \frac{1}{2}(a_n + b_n).$$

Prove that a_n (resp. b_n) are monotonically increasing (resp. decreasing), $a_n < b_n$ and $(b_n - a_n) \to 0$.

7. Let $0 < r < 1$. Show that $\lim_{n \to \infty} r^n = 0$.
8. Set $a_n = 1 + r + \cdots + r^{n-1}$. Show that

$$\lim_{n \to \infty} a_n = \frac{1}{1 - r}$$

provided $|r| < 1$.

9. Show that every rational number can always be expressed as a cyclic decimal.

10. Show that the limit value of a cyclic decimal is always equal to a rational number.

   [Thus irrational numbers are exactly those infinite decimals which are *non-cyclic!*]

---

§ 2. Limit and the Continuity of a Function

Geometrically speaking, a function $y = f(x)$ is continuous if its graph is a continuous curve. Intuitively, it roughly means that the change in the functional values will be small provided the change in the independent variable x is sufficiently small.

A mathematical definition of the localized concept of the continuity of $f(x)$ at $x = a$ was already given in Sec. 2 of Chapter 2. It is convenient to translate that definition in terms of sequential limit, namely,

Definition. A function $f(x)$ is said to be continuous at $x = x_0$ if to any given sequence $\{s_n\}$, $s_n \to x_0$, the corresponding sequence $\{f(s_n)\}$ converges to $f(x_0)$ as its limit, i.e., $f(s_n) \to f(x_0)$.

A function $f(x)$, $a \le x \le b$, is said to be continuous over $[a, b]$ if it is continuous at every point $x_0 \in [a, b]$.

Continuous functions over a closed interval forms an important family of functions in the study of calculus. They enjoy some nice properties which, in fact, plays a basic role in the framework of calculus. The following so-called *intermediate value theorem* is such an example.

Theorem 3.2. *If $f(x)$ is a continuous function defined over $[a, b]$ and v_0 is an arbitrary value between $f(a)$ and $f(b)$, then there exists a suitable place $a \leq x_0 \leq b$ such that $f(x_0) = v_0$.*

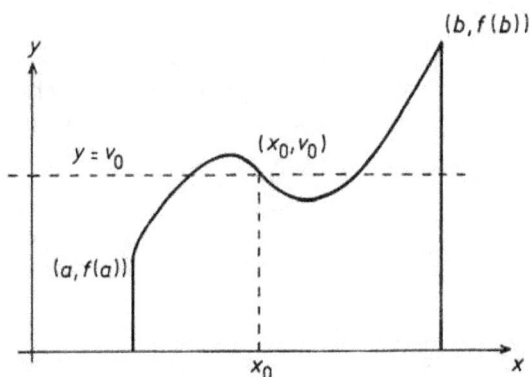

Fig. 24

Analysis

Geometrically, the straight line $y = v_0$ divides the plane into two regions, namely, the region with $y > v_0$ and the region with $y < v_0$. The graph $y = f(x)$ is a continuous curve linking the two points $(a, f(a))$ and $(b, f(b))$ located at different regions separated by the line $y = v_0$. Therefore, it follows from the intuitive meaning of continuity that there must be at least one crossing point, say (x_0, v_0). This is exactly what we are going to prove. From the viewpoint of function, it is not difficult to see that an arbitrary small neighborhood of such a crossing place, say x_0, is characterized by the property:

$$(f(x_0 - \varepsilon) - v_0) \cdot (f(x_0 + \varepsilon) - v_0) < 0$$
(i.e., $(f(x_0 - \varepsilon) - v_0)$ and $(f(x_0 + \varepsilon) - v_0)$ are of different signs).

$$(13)$$

The above analysis clearly suggests the following proof:

Proof. Again, we shall use the method of successive bisection to construct a pair of approach sequences $\{a_n\}$ and $\{b_n\}$ (or equivalently, a nest of intervals $\{[a_n, b_n], \; n \in \mathbb{N}\}$ zero-in onto the yet-to-be proven place of crossing). In view of the above analysis, the guiding criterion for the inductive selection should be

$$(f(a_n) - v_0) \cdot (f(b_n) - v_0) < 0.$$

To be precise, suppose that we have already made the choice of $[a_n, b_n]$ having the above property. We may assume that $f\left(\frac{a_n + b_n}{2}\right) \neq v_0$, otherwise, we may simply take $x_0 = \frac{1}{2}(a_n + b_n)$ and there is nothing left to do. Hence, one and only one of the following two products must be negative, namely

$$(f(a_n) - v_0) \cdot \left(f\left(\frac{1}{2}(a_n + b_n)\right) - v_0 \right);$$

$$\left(f\left(\frac{1}{2}(a_n + b_n)\right) - v_0 \right) (f(b_n) - v_0).$$

Therefore, we simply set $[a_{n+1}, b_{n+1}] = \left[a_n, \frac{1}{2}(a_n + b_n)\right]$ if the first product is negative, otherwise, set $[a_{n+1}, b_{n+1}] = \left[\frac{1}{2}(a_n + b_n), b_n\right]$.

In this way, we construct a pair of approaching sequences $\{a_n\}$ and $\{b_n\}$ such that

$$(f(a_n) - v_0)(f(b_n) - v_0) < 0 \text{ for all } n \in \mathbb{N}. \tag{14}$$

Again, by the continuity of real number system, they determine a common limit $x_0 = \lim_{n \to \infty} a_n = \lim_{n \to \infty} b_n$.

Finally, it follows from the continuity of $f(x)$ at $x = x_0$ that

$$\lim_{n \to \infty} (f(a_n) - v_0) \cdot (f(b_n) - v_0)$$

$$= (\lim_{n \to \infty} f(a_n) - v_0) \cdot (\lim_{n \to \infty} f(b_n) - v_0) \tag{15}$$

$$= (f(x_0) - v_0) \cdot (f(x_0) - v_0).$$

On the other hand, since $(f(a_n) - v_0) \cdot (f(b_n) - v_0) < 0$ for all $n \in \mathbb{N}$, the above limit must be ≤ 0. However, the square of a real number is ≤ 0 only when it is, in fact, equal to 0. This proves that $f(x_0) = v_0$.

$\square$

Exercises

1. Let $a > 1$ be a given real number. Show that there exists a unique positive real number x_0 such that $x_0^n - a = 0$.
   [Hint: Use Theorem 3.2 to show the existence.]
2. The unique positive real root of $(x^n - a)$ is defined to be $\sqrt[n]{a}$ or $a^{\frac{1}{n}}$. Show that $\lim_{n \to \infty} a^{\frac{1}{n}} = 1$.
3. Define $a^{\frac{m}{n}}$ to be equal to $\left(a^{\frac{1}{n}}\right)^m$. Show that

$$\frac{m}{n} < \frac{p}{q} \Rightarrow a^{\frac{m}{n}} < a^{\frac{p}{q}}.$$

4. Let α be an irrational number and $\{r_n\}$ and $\{s_n\}$ be a pair of approach sequence of α, namely

$$r_n \to \alpha \leftarrow s_n .$$

   Show that $\{a^{r_n}\}$ and $\{a^{s_n}\}$ are a pair of approaching sequences.
5. The unique positive real number defined by the above approaching sequences $\{a^{r_n}\}$ and $\{a^{s_n}\}$ is defined to be a^{α}, namely

$$r_n \to \alpha \leftarrow s_n \Leftrightarrow a^{r_n} \to a^{\alpha} \leftarrow a^{s_n}.$$

   Thus a^{α} is now defined for all real numbers α. Show that the function $f(x) = a^x$ is a monotonically increasing continuous function.
6. Let θ be a positive small angle. Show that
   (i) $\frac{1}{2} \sin \theta \cos \theta < \frac{1}{2}\theta < \frac{1}{2} \tan \theta = \frac{1}{2} \frac{\sin \theta}{\cos \theta}$.
   (ii) $\cos \theta < \frac{\sin \theta}{\theta} < \frac{1}{\cos \theta}$.

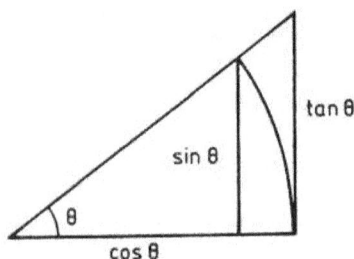

Fig. 25

(iii) $\lim_{\theta \to 0} \frac{\sin \theta}{\theta} = 1$.
[Hint: Fig. 25.]
7. Use Exercise 6(iii) to deduce that
  (i) $\lim_{\theta \to 0} \frac{1 - \cos \theta}{\theta} = 0$.
  (ii) $\lim_{\theta \to 0} \frac{1 - \cos \theta}{\theta^2} = \frac{1}{2}$.

---

§ 3. Approximation and Integration of Piecewise Monotonic Functions

Recall that a function $f(x)$, $x \in [a, b]$, is said to be *piecewise monotonic* if the interval of definition $[a, b]$ can be suitably subdivided into a finite number of subintervals, $\{[a_{i-1}, a_i], \ 1 \le i \le l\}$, such that the restriction of $f(x)$ to each subinterval is either monotonically increasing or monotonically decreasing. In this subsection, we shall use the approximation method to clarify the concept of $\int_a^b f(x)dx$ for such functions, namely, to establish an *analytical definition* of the definite integral of a piecewise monotonic function.

Following the same general idea of Sec. 2.4 of Chapter 2 it is quite natural to use the step functions as our approximating basic functions in the discussion of integration. As it was already pointed out in Sec. 2.4 of Chapter 2 it follows from the intuitive meaning of integration that

(i) $\displaystyle\int_a^b f(x)dx = \sum_{i=1}^{l}\int_{a_{i-1}}^{a_i} f(x)dx,$

$$a = a_0 < a_1 < \cdots < a_i < \cdots < a_l = b, \tag{16}$$

(ii) if $f(x) \le g(x)$ for all $x \in [a, b]$, then

$$\int_a^b f(x)dx \le \int_a^b g(x)dx.$$

Without loss of generality, we may simply assume that $f(x)$, $x \in [a, b]$, is monotonically increasing. For each $n \in \mathbb{N}$, one subdivides the interval $[a, b]$ into $l = 2^n$ subintervals of equal length, namely,

$$a = a_0 < a_1 < a_2 < \cdots < a_{i-1} < a_i < \cdots < a_{l-1} < a_l = b$$

$$\text{and } (a_i - a_{i-1}) = \left(\frac{1}{2}\right)^n \cdot (b - a). \tag{17}$$

Set $G_n(x)$ and $g_n(x)$ to be the following step functions:

$$\begin{cases} G_n(x) = f(a_i) \text{ for } a_{i-1} < x \le a_i \\ g_n(x) = f(a_{i-1}) \text{ for } a_{i-1} \le x < a_i \end{cases}, \ 1 \le i \le l = 2^n. \tag{18}$$

Then, it follows from the increasing property of $f(x)$ that

$$g_1(x) \le g_2(x) \le \cdots \le g_n(x) \le g_{n+1}(x) \le \cdots \le f(x) \le \ldots$$

$$\le G_{n+1}(x) \le G_n(x) \le G_2(x) \le G_1(x) \text{ for all } x \in [a, b]. \tag{19}$$

Moreover,

$$\int_a^b G_n(x)dx = \sum_{i=1}^{l} f(a_i) \cdot (a_i - a_{i-1}) = \frac{(b-a)}{l} \cdot \sum_{i=1}^{l} f(a_i)$$

$$\int_a^b g_n(x)dx = \sum_{i=1}^{l} f(a_{i-1})(a_i - a_{i-1}) = \frac{(b-a)}{l} \cdot \sum_{i=1}^{l} f(a_{i-1}). \tag{20}$$

Therefore,

$$\int_a^b g_1(x)dx \le \int_a^b g_2(x)dx \le \cdots \le \int_a^b g_n(x)dx$$

$$\le \int_a^b g_{n+1}(x)dx \le \cdots \le \int_a^b f(x)dx \le \cdots \le \int_a^b G_{n+1}(x)dx$$

$$\le \int_a^b G_n(x)dx \le \cdots \le \int_a^b G_2(x)dx \le \int_a^b G_1(x)dx \tag{21}$$

and $\displaystyle\int_a^b G_n(x)dx - \int_a^b g_n(x)dx = \frac{(b-a)}{l} \cdot (f(b) - f(a)) \to 0$

as $n \to \infty$.

In other words, $\left\{ \int_a^b g_n(x)dx \right\}$ and $\left\{ \int_a^b G_n(x)dx \right\}$ constitute a pair of approaching sequences, and their common limit should be exactly $\int_a^b f(x)dx$. This clearly provides an analytical definition of $\int_a^b f(x)dx$ that we are seeking.

---

§ 4. Approximation and Differentiation of Piecewise Smooth Function

The concept of the rate of change of a piecewise linear function is extremely elementary. Nevertheless it is undefined at those corner points and it takes constant value on each subinterval between two consecutive corner points. In this section, we shall study the concept of the rate of change of a much more general class of functions which *we shall call piecewise smooth functions*. Geometrically speaking, a function $f(x)$ is said to be *piecewise smooth* if its graph, $y = f(x)$, is piecewise smooth, namely, *locally smooth* except, possibly, a finite number of corner points. Intuitively speaking, the *local smoothness* of a curve γ in the neighborhood of a point $P_0 \in \gamma$ can be described as follows:

If one looks at a small segment of the curve γ in the vicinity of P_0 through microscopes of larger and larger magnification factors, then it looks more and more like a straight line. For example, if one looks at a circle of radius 1 meter, it is obviously curved. But if one magnified such a circle, say by a factor of 10^7, and then look at a segment of length 1 meter.

It will definitely look like a perfect straight line. (It becomes a circle of radius 10^7 meters which is even bigger than the equator of the earth.) Or in other words, the local smoothness of a curve γ at a point $P_0 \in \gamma$ means the *existence of a "tangent line"* of γ at P which, under a high power microscope, looks almost identical to that of a small segment of γ around P_0.

Conceptually, it is quite natural to define the tangent line of γ at P_0 as the limit position of the secant lines $\overline{P_0 P_1}$, $P_0, P_1 \in \gamma$, when $P_1 \to P_0$ as a limit. For example, in the typical case that the curve γ is the graph of $y = f(x)$, $P_0 = (x_0, f(x_0))$, and $P_1 = (x_1, f(x_1))$, the slope of the secant line $\overline{P_0 P_1}$ is equal to

$$\frac{f(x_1) - f(x_0)}{x_1 - x_0} = m = \tan\theta. \tag{22}$$

Therefore, as $P_1 \to P_0$, the secant line $\overline{P_0 P_1}$ approaches a limit position if and only if the above quotient, $\frac{f(x_1) - f(x_0)}{x_1 - x_0}$, has a limit value as $x_1 \to x_0$. Hence, it is quite natural to define the concept of the rate of change as follows:

Definition. If there exists a real number m_0 such that

$$\frac{1}{h}[f(x_0 + h) - f(x_0)] \to m_0 \text{ as } h \to 0, \tag{23}$$

then m_0 is defined to be the rate of change of $f(x)$ at x_0. (In terms of formal logical language, the above limit means that to any given $\varepsilon > 0$, there exists a sufficiently small $\delta > 0$, such that $\left| \frac{1}{h}[f(x_0 + h) - f(x_0)] - m_0 \right| < \varepsilon$ provided $|h| < \delta$.) Geometrically,

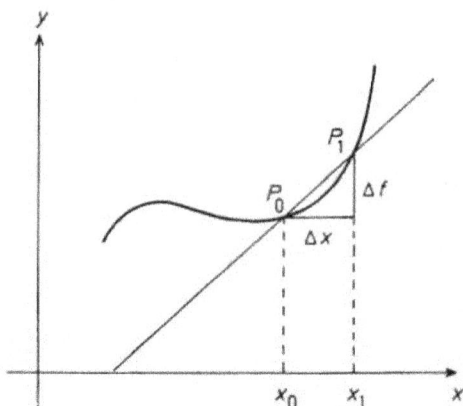

Fig. 26

m_0 is the slope of the tangent line of the graph $y = f(x)$ at the point $P_0(x_0, f(x_0))$, namely, the equation of the tangent line at P_0 is given by: $y - f(x_0) = m_0(x - x_0)$. Hence, the *local smoothness* of the graph $y = f(x)$ at the point $P_0(x_0, f(x_0))$ is equivalent to the existence of the limit of the above quotient.

Example 1. $y = x^n$, x_0 arbitrary

$$\frac{1}{h}[(x_0 + h)^n - x_0^n]$$
$$= \frac{1}{h}\left[\left(x_0^n + nx_0^{n-1} \cdot h + \frac{n(n-1)}{2}x_0^{n-2} + \cdots + h^n\right) - x_0^n\right] \quad (24)$$
$$= nx_0^{n-1} + \frac{n(n-1)}{2}x_0^{n-2} \cdot h + \cdots + h^{n-1}.$$

Observe that the first term is the only term without a factor of h, therefore, as $h \to 0$, $\frac{1}{h}[(x_0 + h)^n - x_0^n] \to nx_0^{n-1}$.

Example 2. $y = \sin x$, $x_0 = 0$

$$\frac{1}{h}[\sin h - \sin 0] = \frac{\sin h}{h} = \frac{2\sin h}{2h}. \quad (25)$$

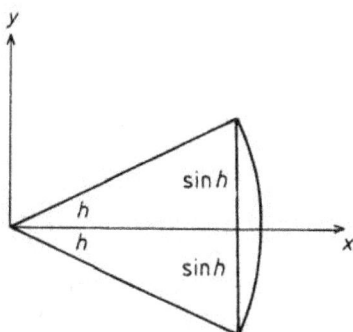

Fig. 27

Geometrically, $2 \sin h$ is the length of the cord $\overline{AA'}$ and $2h$ is the length of the circular arc, $\overset{\frown}{AA'}$. Therefore, it is intuitively clear that their ratio $\to 1$ as $h \to 0$.

Example 3. $y = \sin x$, x_0 arbitrary. We need to find out the limit value of the following quotient:

$$\frac{1}{h}[\sin(x_0 + h) - \sin x_0] = \frac{1}{h}[\sin x_0 \cos h + \cos x_0 \sin h - \sin x_0] \tag{26}$$
$$= \sin x_0 \cdot \frac{\cos h - 1}{h} + \cos x_0 \cdot \frac{\sin h}{h}.$$

By Example 2 $\frac{\sin h}{h} \to 1$ as $h \to 0$. Based on that, it is not difficult to see that, as $h \to 0$,

$$\frac{\cos h - 1}{h} = \frac{(\cos^2 h - 1)}{h(\cos h + 1)} = -\frac{\sin h}{h} \cdot \sin h \cdot \frac{1}{\cos h + 1} \to 1 \cdot 0 \cdot \frac{1}{2} = 0. \tag{27}$$

Hence, as $h \to 0$,

$$\frac{1}{h}[\sin(x_0 + h) - \sin x_0] \to \sin x_0 \cdot 0 + \cos x_0 \cdot 1 = \cos x_0. \tag{28}$$

That is, the rate of change of $\sin x$ at x_0 is equal to $\cos x_0$.

Exercises

A subset $S \subset \mathbb{R}$ is said to be bounded above (resp. below) if there exists a number K such that

$$s \leq K \text{ (resp. } s \geq K)$$

for all $s \in S$. Such a K is called an upper (resp. lower) bound of S.

1. Show that a subset $S \subset \mathbb{R}$ which is bounded above (resp. below) always has a *least upper bound* (resp. *greatest lower bound*). [Hint: Use the same kind of proof as that of Theorem 3.1.]

2. Let $f(x)$ be a cubic polynomial of real coefficients. Show that it has at least one real root. [Hint: Show that $f(K)$ and $f(-K)$ are necessarily of opposite signs for sufficiently large K and then apply Theorem 3.2.]

3. Show that any real coefficient polynomial of odd degree always has at least one real root.

   A function $f(x)$ is said to be *strictly* monotonically increasing (resp. decreasing) if

   $$x_1 > x_2 \Rightarrow f(x_1) > f(x_2) \text{ (resp. } f(x_1) < f(x_2)).$$

   If the relationship between a pair of variables x and y is such that they are functions of each other, namely $y = f(x)$ and $x = g(y)$, then g is called the inverse function of f and vice versa.

4. If $y = f(x)$ is a *continuous and strictly monotonically increasing* (resp. *decreasing*) function defined on $[a, b]$, then there exists an *inverse function* $x = g(y)$ defined on $[f(a), f(b)]$ (resp. $[f(b), f(a)]$) which is also continuous and strictly monotonically increasing (resp. decreasing). [Hint: Use Theorem 3.2 for the existence of x_0 for each $y_0 \in [f(a), f(b)]$ (resp. $[f(b), f(a)]$) such that $f(x_0) = y_0$ and use the strict monotonicity to show the uniqueness of such an x_0.]

5. Use the above result of Exercise 4 to investigate the concept of inverse functions for the following ones:
   (i) $y = x^2$
   (ii) $y = \tan x$
   (iii) $y = \sin x$
   (iv) $y = \cos x$.

CHAPTER 4

Foundational Framework and Fundamental Theory of Calculus

In this chapter we shall summarize and combine the results of the previous three chapters to build the foundational framework and the fundamental theory of calculus. Generally speaking, calculus is a branch of mathematics which provides the general framework as well as the basic tools for the quantitative analysis of problems or systems of variable quantities. In most cases, the variable quantities are of the measurement type, e.g., length, area, weight, time, etc. The mathematical system that one uses to represent and to compute the quantitative aspect of these types of quantities is the *real number system*, $\mathbb{R}$. Hence, the concept of real numbers and the basic properties of the real number system, $\mathbb{R}$, forms the very foundation of calculus.

Mathematically, one uses symbols such as x, y, z, s, t, etc. to represent the values of variable quantities (such symbols are called *"variables"*) and uses *functions* to describe those *correlations* amongst the variable quantities (cf. Sec. 2, Chapter 1). Therefore, *functions* are exactly the primary subject of investigation in calculus. Among the basic properties of functions, the most fundamental and also the most useful ones are the *monotonicity*, the *continuity*, the *rate of change* and the *sum of total effect* that we discussed in Chapter 2. Therefore, the first order of fundamental theory of calculus is the *analytical formulation* (i.e., definition) of the above basic properties, thus *transforming* the *intuitive contents* of such basic properties into *precise mathematical concepts.* It is at this junction that the *method of approximation* and the concept of *limit* become indispensable (cf. Chapter 3).

---

§ 1. The Foundation of Calculus

1.1. *The real number system*

Historically, the debut of the real number system in mathematics occurred in the foundation theory of quantitative geometry. The discovery of *non-commensurable pairs of intervals* by Hippasus in the 5th century B.C. initially posed a detrimental threat to the entire foundation theory of geometry built upon the "universal validity" of commensurability. However, such a fundamental difficulty inspired Eudoxus in the 4th century B.C. to invent the *approximation methodology* which not only enabled him to overcome the difficulty posed by non-commensurable intervals and rebuilt a sound foundation theory of geometry, but it also provides one of the most powerful methodologies that the entire modern mathematical analysis is based upon. Even today, revisiting and reconstructing the above remarkable historical development remains to be the most natural starting point of understanding the real number system and learning calculus.

The real number system, $\mathbb{R}$, contains the rational number system, $\mathbb{Q}$, because quantities of the measurement type are infinitely divisible. However, it also contains a lot more irrational numbers corresponding to non-commensurable pairs. Eudoxus taught us that every irrational number can be approximated by rational numbers. For example, a given irrational number α (such as $\frac{1}{2}(1 + \sqrt{5})$ or $\sqrt{2}$) can be characterized by a pair of approaching sequences $\{a_n\}$ and $\{b_n\}$, namely

$$a_1 \leq a_2 \leq \cdots \leq a_n \leq a_{n+1} \leq \alpha \leq b_{n+1} \leq b_n \leq \cdots \leq b_2 \leq b_1$$
$$\text{and } (b_n - a_n) \to 0 \tag{1}$$

or simply represent the above relationship by

$$a_n \rightarrow \alpha \leftarrow b_n.$$

Recall that the *uniqueness aspect* of the above approaching pair of sequences, namely

$$a_n \rightarrow \left\{ \begin{matrix} \alpha \\ \alpha' \end{matrix} \right\} \leftarrow b_n \Rightarrow \alpha = \alpha' \tag{2}$$

is exactly a modern reformulation of the Eudoxusian principle. On the other hand, the *existence of a common limit* for any given pair of approaching sequences is exactly the *analytical formulation* of the *continuity of the real line*, namely

The continuity (or completeness) of the real number system. To any given pair of approaching sequences

$$a_1 \leq a_2 \leq \cdots \leq a_n \leq a_{n+1} \leq b_{n+1} \leq b_n \leq \cdots \leq b_2 \leq b_1$$
$$\text{and } (b_n - a_n) \rightarrow 0 \tag{3}$$

there always exists a (unique) common limit value α, namely $a_n \rightarrow \alpha \leftarrow b_n$.

This is the fundamental existence statement that the proofs of all the other existence theorems in analysis are based upon. For example, the existence of the limit of a bounded monotonic sequence, the intermediate value theorem for continuous functions etc. (cf. Sec. 3.2).

Remarks

(i) Algebraically, the real number system, $\mathbb{R}$, and the rational number system, $\mathbb{Q}$, have the same kind of formal properties such as the commutative and associative laws of both addition and multiplication and the distributive law: $a \cdot (b + c) = a \cdot b + a \cdot c$. The major difference between $\mathbb{R}$ and $\mathbb{Q}$ lies in the aspect of limit, namely, the completeness of the former while the latter is far from being complete, namely, the limit of a converging sequence of rational numbers may very well be an irrational number!

(ii) In the setting of decimal representation, every real number α can be characterized by a pair of approaching sequences $\{a_n\}$ and $\{b_n\}$ where a_n is the unique maximal decimal with n digits which is at most equal to α and $b_n = a_n + \left(\frac{1}{10}\right)^n$. It can be shown that α is a rational number if and only if its decimal representation is *cyclic*. Namely, the irrational numbers are exactly those real numbers whose decimal representations are *non-cyclic*.

1.2. *Approximation and limit*

The basic idea of the approximation methodology is to use some better known, more elementary type of objects to approximate a comparatively more advanced type of object whose understanding is something yet to be achieved! The key point of approximation is that the *difference* between the approximating objects and the target object (measured in a specific quantitative way) can be made as small as one wants. For example, the first resounding success of approximation methodology was the Eudoxian principle which uses the *rationals* to approximate the *irrationals* (cf. Sec. 1.1); one uses the *piecewise linear functions* to approximate the *piecewise smooth functions* in the study of rate of change, while in the study of integration (i.e., the sum of total effect) one uses the *step functions* to approximate the *piecewise* monotonic functions (cf. Sec. 3.2).

Notice that an approximation procedure usually consists of an infinite number of approximating stages. Therefore, corresponding to a given approximation procedure, there is a sequence of objects (such as numbers or functions of a given type) whose n-th term is exactly the n-th stage approximating object. Thus it naturally arises the terminology of sequences and the *concept of limit*.

Logically speaking, to say that *the sequence* $\{s_n\}$ *approximates* α and to say that α *is equal to the limit of* $\{s_n\}$ are simply two equivalent ways of describing the same relationships between $\{s_n\}$ and α.

However, as it has already been pointed out in Chapter 3, the viewpoints and the emphasis of the *approximation* and that of the *limit* are quite *different!* In fact, it is often helpful to bear in mind the following differences between them in the actual usage of approximation and limit, namely

(i) In case the target object is given beforehand, then one takes the *approximation viewpoint* and the emphasis is on the *uniqueness* aspect.

(ii) In case the sequence is given beforehand, then one takes the *limit viewpoint* and the emphasis is on the *existence* aspect.

We include here the following basic existence theorem on limit of sequences which is closely related to the completeness (i.e., continuity) of real number systems, namely

Theorem 4.1 (Cauchy Criterion). *The following Cauchy condition is a necessary and sufficient condition for the existence of limit of a given sequence* $\{a_n\}$, *namely: to any given* $\varepsilon > 0$, *there exists a sufficiently large* N *such that* $|a_m - a_n| < \varepsilon$ *provided* $m, n \geq N$.

Proof. **Necessity:** Suppose that the limit of $\{a_n\}$ exists and α is the limit value. Then, to a given $\varepsilon > 0$, there exists a sufficiently large N such that

$$|a_m - \alpha| < \frac{\varepsilon}{2} \text{ and } |a_n - \alpha| < \frac{\varepsilon}{2} \tag{4}$$

provided $m, n \geq N$. Hence

$$|a_m - a_n| \leq |a_m - \alpha| + |a_n - \alpha| < \frac{\varepsilon}{2} + \frac{\varepsilon}{2} = \varepsilon. \tag{5}$$

Sufficiency: Again, we shall use the method of successive bisection to construct a pair of approaching sequence with a specific point as their common limit. Set N_0 to be a sufficiently large number such that

$$|a_{N_0} - a_n| < 1 \text{ for } n > N_0 \tag{6}$$

and

$$K = \max\{|a_i|, \ 1 \le i \le N_0 - 1, \ |a_{N_0}| + 1\}. \tag{6'}$$

Then all a_n lie inside the interval $[-K, K]$.

Observe that if $[A_k, B_k]$ is a given interval which contains a_n for *infinitely many indices* n, then at least one of its half intervals also contains a_n for infinitely many indices n. Therefore, we can begin with $[A_1, B_1] = [-K, K]$ and inductively choosing $[A_{k+1}, B_{k+1}]$ as one of the half intervals of $[A_k, B_k]$ *preserving the above property of containing a_n for infinitely many indices* n. Of course, it is possible to have both half intervals with the above properties. At such an occasion, one has a choice to make. Suppose that one *always* chooses the right (resp. left) half at such an occasion. Then one obtains a pair of approaching sequences $\{A_k\}$ and $\{B_k\}$ (resp. $\{A_k'\}$ and $\{B_k'\}$) satisfying the following properties, namely

(i) each $[A_k, B_k]$ (resp. $[A_k', B_k']$) contains a_n for infinitely many indices n,

(ii) there are only finitely many indices n with a_n larger (resp. smaller) than B_k (resp. A_k'). Let α (resp. α') be the common limit of $\{A_k\}$ and $\{B_k\}$ (resp. $\{A_k'\}$ and $\{B_k'\}$) whose existence follows from the completeness of the real number system.

Now, we shall make use of the Cauchy's condition to show that $\alpha = \alpha'$. Suppose the contrary that $\alpha \ne \alpha'$. Then it is easy to see that $\alpha > \alpha'$. Set $\varepsilon = \frac{1}{3}(\alpha - \alpha')$. Notice that

$$A_k \to \alpha \leftarrow B_k, \ A_k' \to \alpha' \leftarrow B_k' \tag{7}$$

and hence there exists a sufficiently large k_0 such that

$$\begin{aligned}
[A_{k_0}, B_{k_0}] &\subset [\alpha - \varepsilon, \alpha + \varepsilon] \\
[A_{k_0}', B_{k_0}'] &\subset [\alpha' - \varepsilon, \alpha' + \varepsilon].
\end{aligned} \tag{8}$$

Therefore, no matter how large N is, there always exists $m, n \ge N$ such that $a_m \in [A_{k_0}, B_{k_0}]$ and $a_n \in [A_{k_0}', B_{k_0}']$, thus $a_m - a_n > \varepsilon$,

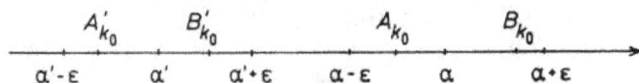

Fig. 28

which contradicts the assumption that $\{a_n\}$ satisfies the Cauchy's condition (cf. Fig. 28).

Hence $\alpha = \alpha'$, and it is not difficult to show that $\alpha = \alpha'$ is, in fact, the limit of the sequence $\{a_n\}$, because there are only finitely many indices n with a_n outside of the ε-neighborhood of $\alpha = \alpha'$.

□

1.3. *Continuous functions*

Among all kinds of variable systems and changing phenomena, sudden abrupt changes are exceptional and only occur at rare, special occasions, namely, in most systems and phenomena in nature gradual continuous changes are, by far, the usual case and for most of the time. Therefore, the functions that represent the correlations among the variable quantities are mostly continuous functions or functions with at most some isolated discontinuity. Intuitively speaking, a function $y = f(x)$, $x \in [a, b]$, is everywhere continuous if its graph is a continuous curve. The analytical definition of the localized continuity of $y = f(x)$ at a given point x_0 can also be reformulated in terms of sequential limit as follows, namely, to each sequence $a_n \to x_0$, the corresponding sequence of functional values $f(a_n) \to f(x_0)$:

$$\lim_{n \to \infty} f(a_n) = f(\lim_{n \to \infty} a_n)$$

provided $\lim_{n \to \infty} a_n$ exists and within the domain of definition of $y = f(x)$.

Continuous functions defined on a *close* interval $[a, b]$ have many useful, nice properties and such basic properties play an important role in the fundamental theory of calculus. For example,

Theorem 4.2. *Let $y = f(x)$ be a continuous function defined on a close interval $[a, b]$. Then there exist $\alpha, \beta \in [a, b]$ such that*

$$f(\alpha) \le f(x) \le f(\beta) \tag{9}$$

for all $x \in [a, b]$.

Remark. Together with the intermediate value theorem, the above theorem shows that the *image set* of such a continuous function on a close interval is again a close interval, namely $[f(\alpha), f(\beta)]$.

Proof. First, one shows that the set of functional values, i.e., $\{f(x) \mid x \in [a, b]\}$ is bounded, namely, $|f(x)| < K$ for a sufficiently large K. Suppose the contrary that it is unbounded. Then one can begin with $[a_1, b_1] = [a, b]$ and inductively choose $[a_{n+1}, b_{n+1}]$ as one of the half interval of $[a_n, b_n]$ such that the functional values of $f(x)$ on $[a_{n+1}, b_{n+1}]$ *remains* unbounded, thus obtaining a pair of approaching sequences $\{a_n\}$ and $\{b_n\}$ with a common limit α inside of $[a, b]$. On the other hand, the continuity of $f(x)$ at $x = \alpha$ implies that the value of $f(x)$ in a sufficiently small neighborhood of α in $[a, b]$, which clearly contains $[a_n, b_n]$ for sufficiently large n, must be bounded. This is, of course, a contradiction.

Set M (resp. m) to be the least upper bound (resp. greatest lower bound) of the bounded set $\{f(x) \mid x \in [a, b]\}$. What we need to show is that both M and m can be realized as functional values of $f(x)$, namely, there exist $\alpha, \beta \in [a, b]$ such that $f(\alpha) = m$ and $f(\beta) = M$. We shall again prove the existence of such an α (resp. β) by constructing a pair of approaching sequences converging to it. Again begin with $[a_1, b_1]$ (resp. $[a'_1, b'_1]) = [a, b]$ and inductively choose $[a_{n+1}, b_{n+1}]$ (resp. $[a'_{n+1}, b'_{n+1}]$) as one of the half intervals of $[a_n, b_n]$ (resp. $[a'_n, b'_n]$) such that the least upper bound (resp. greatest lower bound) of the functional values of $f(x)$ over $[a_{n+1}, b_{n+1}]$ (resp. $[a'_{n+1}, b'_{n+1}]$) *remains to be* equal to M (resp. m). Let β (resp. α) be their common limit, namely

$$a_n \to \beta \leftarrow b_n \text{ (resp. } a'_n \to \alpha \leftarrow b'_n). \tag{10}$$

Then, it is easy to see that $f(\beta) = M$ (resp. $f(\alpha) = m$). For otherwise, $f(\beta) < M$ (resp. $f(\alpha) > m$) and it follows from the continuity of $f(x)$ at $x = \beta$ (resp. $x = \alpha$) that the least upper bound (resp. greatest lower bound) of the functional values of $f(x)$ over a sufficiently small neighborhood of β (resp. α) is strictly smaller (resp. larger) than M (resp. m). This is a contradiction to the above construction because $[a_n, b_n]$ (resp. $[a'_n, b'_n]$) are contained in the above small neighborhood of β (resp. α) for sufficiently large n and the least upper bounds (resp. greatest lower bounds) of the functional values over $[a_n, b_n]$ (resp. $[a'_n, b'_n]$) *remain* to be equal to M (resp. m) for *all* n.

1.4. *Sum of Total Effect, Area and Integration*

Intuitively speaking, suppose that $f(t)$ records the rate of a certain process as a function of time such as speed, rate of flow or rate of electricity consumption, etc. Then the *sum of total effect* over a time interval of $[a, b]$ is called the definite integral of $f(t)$ over $[a, b]$ and it is denoted by $\int_a^b f(t)dt$. Geometrically, the definite integral $\int_a^b f(t)dt$ can be interpreted as the oriented area of the region bounded by the graph of $y = f(x)$, the x-axis and the two vertical bounds $x = a$ and $x = b$, as indicated in Fig. 29.

For the study of integration, the family of step functions are both elementary and basic. On the one hand, the definite integral of a step function can easily be expressed as a simple sum, namely

$$\int_a^b g(x)dx = \sum_{i=1}^l \int_{a_{i-1}}^{a_i} g(x)dx = \sum_{i=1}^l c_i(a_i - a_{i-1}) \tag{11}$$

where $g(x) = c_i$ for $a_{i-1} < x < a_i$. On the other hand, to any given piecewise monotonic (or continuous) function $f(x)$ defined on a close interval $[a, b]$, a suitable pair of approaching sequence of step functions $\{g_n(x)\}$ and $\{G_n(x)\}$ can be constructed such that

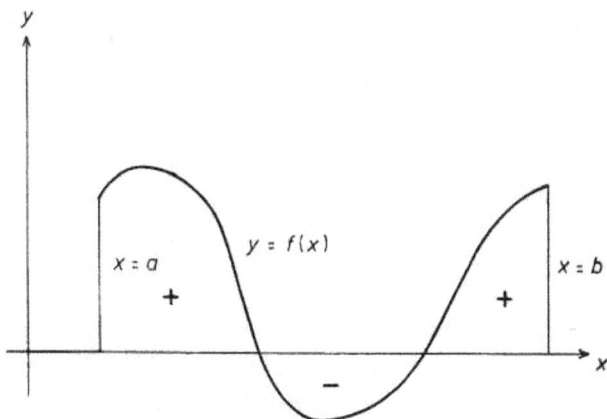

Fig. 29

$$g_n(x) \leq f(x) \leq G_n(x), \ a \leq x \leq b \tag{12}$$

and

$$\int_a^b g_n(x)dx \rightarrow \int_a^b f(x)dx \leftarrow \int_a^b G_n(x)dx \ , \tag{12'}$$

thus characterizing the integral $\int_a^b f(x)dx$ as the common limit of the two approaching sequences of numbers, namely $\{\int_a^b g_n(x)dx\}$ and $\{\int_a^b G_n(x)dx\}$. This is exactly the analytical formulation (or rather, definition) of the definite integral $\int_a^b f(x)dx$.

Based upon the above analytical definition of the definite integrals, it is not difficult to verify the following basic properties:

(i) $\int_a^b f(x)dx + \int_b^c f(x)dx = \int_a^c f(x)dx$.

(ii) If $g(x) \leq f(x) \leq G(x)$ for all $x \in [a,b]$, then

$$\int_a^b g(x)dx \leq \int_a^b f(x)dx \leq \int_a^b G(x)dx.$$

(iii) $\int_a^b [f_1(x) + f_2(x)]dx = \int_a^b f_1(x)dx + \int_a^b f_2(x)dx$.

(iv) $\int_a^b c \cdot f(x)dx = c \cdot \int_a^b f(x)dx$, c is a constant.

Theorem 4.3 (Integral Mean Value Theorem). *If $f(x)$ is a continuous function defined on the close interval $[a, b]$, then there exists a suitable point $x_0 \in [a, b]$ such that*

$$\int_a^b f(x)dx = f(x_0) \cdot (b - a). \tag{13}$$

(Geometrically, the above equation means that the area of the curved region is equal to the area of the rectangle of height $f(x_0)$ and width $(b - a)$. Functionally, it means that $f(x_0)$ is the mean value of $f(x)$ over the whole interval $[a, b]$.)

Proof. By Theorem 4.2, there exist $\alpha, \beta \in [a, b]$ such that

$$m = f(\alpha) \leq f(x) \leq f(\beta) = M, \ a \leq x \leq b. \tag{14}$$

Therefore, by (ii)

$$m(b - a) = \int_a^b m\,dx \leq \int_a^b f(x)dx \leq \int_a^b M \cdot dx = M(b - a) \tag{15}$$

namely

$$f(\alpha) = m \leq \frac{1}{b - a} \int_a^b f(x)dx \leq M = f(\beta). \tag{16}$$

Hence, by the intermediate value theorem of Chapter 3, there exists a suitable point $x_0 \in [a, b]$ such that

$$\frac{1}{b - a} \int_a^b f(x)dx = f(x_0), \ \text{i.e.,} \ \int_a^b f(x)dx = f(x_0) \cdot (b - a). \tag{17}$$

$\square$

1.5. *Rate of change, slope and differentiation*

Recall that the *rate of change* is a *local property*, namely, the rate of change of a function $f(x)$ at $x = x_0$ only depends on the functional values of $f(x)$ restricted to an arbitrary small neighborhood of x_0. Associated to a given function $f(x)$, there is in general another function which records the rate of change of $f(x)$ at each individual point $x_0 \in [a, b]$. Such a function is, nowadays, called the *derivative* of $f(x)$ and is usually denoted by $f'(x)$, namely, $f'(x_0)$ is, by definition, equal to the rate of change of $f(x)$ at $x = x_0$ (if it can be properly defined).

Geometrically, $f'(x_0)$ is equal to the slope of the tangent line of the graph $y = f(x)$ at $P(x_0, f(x_0))$ while the tangent line is defined to be the limit line of the secant line $P_0 P_1$ as $P_1 \to P_0$ along the graph (as indicated in Fig. 30).

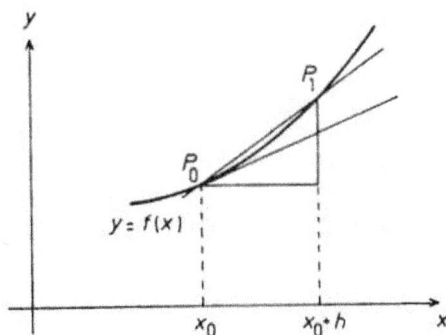

Fig. 30

Analytically, $f'(x_0)$ can be simply defined as the limit of the quotient, $\frac{1}{h}[f(x_0 + h) - f(x_0)]$, which is usually called the average rate of change of $f(x)$ over the interval $[x_0, x_0 + h]$ (resp. $[x_0 + h, x_0]$ if h is negative), namely

$$f'(x_0) = \lim_{h \to 0} \frac{1}{h}[f(x_0 + h) - f(x_0)]. \tag{18}$$

(Of course, $f'(x_0)$ is defined only when the above limit actually exists.)

Examples

(1) A *piecewise linear* function $f(x)$ is characterized by the property that its rates of change are locally constant, namely, its derivative $f'(x)$ is a step function.

(2) If $f(x) = x^n$, then $f'(x) = nx^{n-1}$.

(3) If $f(x) = \sin x$, then $f'(x) = \cos x$. We shall consider the computation of the derivative, $f'(x)$, from a given function, $f(x)$, as an operation among functions and call such an operation *differentiation*. It is not difficult to verify the following simple but useful rules of differentiation:

(i) $[f_1(x) + f_2(x)]' = f_1'(x) + f_2'(x)$.

(ii) $[c \cdot f(x)]' = c \cdot f'(x)$.

(iii) $[f(x) \cdot g(x)]' = f'(x) \cdot g(x) + f(x) \cdot g'(x)$.

(iv) $\left[\frac{1}{f(x)}\right]' = \frac{-f'(x)}{[f(x)]^2}$.

(v) $f[g(x)]' = f'[g(x)] \cdot g'(x)$

(chain rule for differentiation of composite functions).

In general, differentiation is technically a rather straightforward operation.

---

§ 2. Fundamental Theory of Calculus

2.1. *The relationship between integration and differentiation —the fundamental theorem of analysis*

The "sum of total effect" and the "rate of change" are clearly two basic quantitative properties of functions with fundamental importance. However, they are quite different in nature; the former is a *global* property which depends on the functional values over the whole interval but the latter is a *local* property which is defined pointwise and only depends on the functional values of an arbitrarily small neighborhood of that point. Corresponding to the above two basic quantitative properties of functions, one has two fundamental operations of calculus, namely, integration and differentiation. The former computes a definitive number called the definite integral of a given function $f(x)$ over a given interval $[a, b]$, $\int_a^b f(x)dx$, while the latter computes a function called the derivative of a given function $f(x)$, denoted by $f'(x)$. Of course, one naturally expects that both of them will play important roles in calculus. In fact, the term "*calculus*" itself is exactly the abbreviation of "*integral and differential calculus*". However, it is such a wonderful internal harmony that these two rather different fundamental operations are actually neatly and tightly related! This powerful interlocking relationship between the two foundational operations is exactly the conclusion of the *fundamental theorem of calculus*.

Let us first begin with some simple but relevant observations, namely

(i) Since the integral over the whole interval $[a, b]$, $\int_a^b f(t)dt$, is a number; the integral over a partial interval $[a, x] \subset [a, b]$, i.e., $\int_a^x f(t)dt$, is of course a number whose value depends on the end point x. Therefore

$$F(x) = \int_a^x f(t)dt \qquad (19)$$

is, in fact, a function of x which is obtained from the given function $f(x)$ via integration.

(ii) Let $f(x)$ be a step function defined on $[a, b]$ and c_i be the constant value of $f(x)$ over the i-th subinterval (a_{i-1}, a_i). Then, it is easy to see that

$$F(x) = \int_a^x f(t)dt, \ a < x \leq b$$

is a piecewise linear function whose graph over the i-th subinterval $[a_{i-1}, a_i]$ is a straight segment with c_i as its slope. Hence the rate of change of $F(x)$ at $x_0 \in (a_{i-1}, a_i)$ is exactly c_i, namely, $F'(x)$ is just the original step function $f(x)$!

The above simple basic facts and the relationships amongst many concrete examples such as the one on speed and mileage naturally suggest the general validity of the following theorem, namely

Fundamental Theorem of Analysis. *Let $f(x)$ be a continuous function defined on a close interval $[a, b]$ and set*

$$F(x) = \int_a^x f(t)dt. \tag{19}$$

Then $F'(x) = f(x)$ for all $a < x < b$.

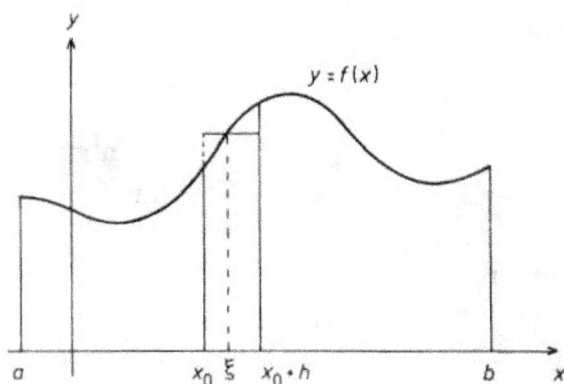

Fig. 31

Proof. By the above definition and Theorem 4.3

$$\frac{1}{h}[F(x_0 + h) - F(x_0)] = \frac{1}{h}\left\{\int_a^{x_0+h} f(t)dt - \int_a^{x_0} f(t)dt\right\}$$

$$= \frac{1}{h}\int_{x_0}^{x_0+h} f(t)dt = f(\xi) \tag{20}$$

for a suitable ξ lies between x_0 and $x_0 + h$. Now, let $h \to 0$ (i.e., $x_0 + h \to x_0$). Then ξ, being sandwiched between x_0 and $x_0 + h$, has nowhere to go but tends to x_0 as a limit! Hence, it follows from the continuity assumption of $f(x)$ that $f(\xi) \to f(x_0)$ as $\xi \to x_0$. This proves that

$$\lim_{h \to 0} \frac{1}{h}[F(x_0 + h) - F(x_0)] = f(x_0) \qquad (21)$$

namely $F'(x_0) = f(x_0)$. □

2.2. Mean value theorems

The integral version of mean value theorem has already been proved in Sec. 1.4, Chapter 4. If one combines it with the above fundamental theorem of analysis, it can be translated into the following differential version of mean value theorem, namely

Theorem 4.4. *If* $F(x)$, $x \in [a, b]$, *is a function whose derivative* $F'(x)$ *is defined and continuous for all* $x \in [a, b]$, *then*

$$F(b) - F(a) = F'(x_0) \cdot (b - a) \qquad (22)$$

for a suitable $a < x_0 < b$.

Proof. Set $F'(x) = f(x)$ and

$$G(x) = \int_a^x f(t)dt. \qquad (23)$$

Then, by the fundamental theorem of analysis

$$[F(x) - G(x)]' = F'(x) - G'(x) = f(x) - f(x) \equiv 0. \qquad (24)$$

Therefore, $F(x) - G(x)$ is a function with identically zero rate of change and hence is a constant function. That is,

$$F(x) - G(x) \equiv F(a) - G(a) = F(a) \text{ or } F(x) = G(x) + F(a). \qquad (25)$$

Thus

$$F(b) - F(a) = G(b) = \int_a^b f(t)dt = f(x_0) \cdot (b - a)$$
$$= F'(x_0) \cdot (b - a)$$

(26)

for a suitable $a < x_0 < b$. ☐

Remarks

(i) Geometrically, Theorem 4.4 means that there exists a suitable point $P_0(x_0, F(x_0))$ on the graph $y = F(x)$ such that the tangent line at P_0 is parallel to the secant line linking $A(a, F(a))$ to $B(b, F(b))$, as indicated in Fig. 32.

(ii) It is not difficult to give a direct proof of Theorem 4.4 even under a slightly weaker condition, namely

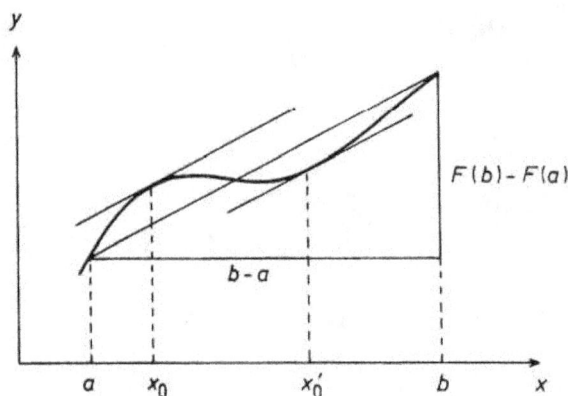

Fig. 32

Theorem 4.4'. *Let $F(x)$ be continuous on the close interval $[a, b]$ and differentiable on the open interval (a, b). Then, there exists a suitable point $a < x_0 < b$ such that*

$$F(b) - F(a) = F'(x_0) \cdot (b - a).$$

(27)

Proof. Intuitively, a point $P_0(x_0, F(x_0))$ on the graph of $y = F(x)$ with maximal distance to the secant line AB should be such a point with its tangent line parallel to AB. It is well known that the equation of the secant line AB is given by

$$[F(b) - F(a)](X - a) - (b - a)[Y - F(a)] = 0 \qquad (28)$$

and the distance between the moving point $P(x, F(x))$ on the graph and the line AB is given by

$$\frac{1}{K}\{[F(b) - F(a)](x - a) - (b - a)[F(x) - F(a)]\} \qquad (29)$$

where $K = \pm[(F(b)-F(a))^2+(b-a)^2]^{1/2}$ is a constant. This motivates us to introduce the following auxiliary function, namely, set

$$G(x) = [F(b) - F(a)](x - a) - (b - a)[F(x) - F(a)]. \qquad (30)$$

It is also continuous on the close interval $[a, b]$ and differentiable on the open interval, and moreover, $G(a) = G(b) = 0$. Hence, by Theorem 4.2, there exist x_0 and x_0' in $[a, b]$ such that

$$m = G(x_0) \leq G(x) \leq G(x_0') = M \qquad (31)$$

for all $a \leq x \leq b$. If $m = M = 0$, then $G(x) \equiv 0$ on $[a, b]$ and $F'(x) \equiv [F(b) - F(a)]/(b - a)$. Therefore, one needs only to consider the case that at least one of m and M is non-zero. Hence, at least one of x_0 and x_0' must be an interior point of $[a, b]$ thus with vanishing derivative of $G(x)$. This proves the existence of an interior point x_0 with $F'(x_0) = (F(b) - F(a))/(b - a)$. □

Remark. Technically, the above mean-value theorem is very handy and useful because it establishes an equality between the finite quotient, $(F(b) - F(a))/(b - a)$, and the derivative function which is easier to perform algebraic manipulations. It is convenient to use because it often enables us to provide estimates or to delay the limiting procedure until the quantities involved have already been transformed into a simpler form (cf. Chapter 5 for some typical examples of such applications of the above mean value theorem).

2.3. *Differentiation and integration, basic laws of operation and a convenient system of notations*

Real number system, variables and functions are the basic framework of mathematical analysis; the *rate of change* and the *sum of total effect* are the two basic quantitative properties of functions whose *analytical formulations* are themselves the very basic part of the foundational theory of calculus, thus defining the two *fundamental operations* among functions, namely, *differentiation* and *integration*. The former computes the *derivative* of a given function $f(x)$, usually denoted by $f'(x)$, which records the rate of change of $f(x)$ at those differentiable points while the latter provides another function, namely

$$F(x) = \int_a^x f(t)dt \tag{19}$$

which records the definite integral (sum of total effect) of $f(t)$ over the interval $[a, x]$. The fundamental theorem of analysis proves that the above two operations are, in fact, reciprocal to each other, namely

$$F'(x) = f(x).$$

For this reason, it is natural and convenient to introduce the following definition and notation, namely

Definition. $F(x)$ is called an *indefinite integral* (or an *antiderivative*) of $f(x)$ if $F'(x) = f(x)$. We shall denote an indefinite integral of $f(x)$ by $\int f(x)dx$.

Remark. Two indefinite integrals of a given function $f(x)$ are differed by a constant, namely

$$F_1'(x) = F_2'(x) = f(x) \Leftrightarrow F_1(x) - F_2(x) \equiv \text{constant}.$$

Before we proceed to discuss the basic laws of differentiation and integration, let us first introduce the system of *differential notations* designed by Leibnitz, one of the co-founders of calculus.

Let $y = f(x)$ be a given function, $\Delta x = x - x_0$ be an *increment* of x and $\Delta y = f(x) - f(x_0)$ be the corresponding increment of y. Recall that the rate of change of $f(x)$ at x_0 is defined to be the limit of the quotient $\Delta y / \Delta x$ (if such a limit exists). Leibnitz used a rather suggestive notation $\frac{dy}{dx}$ to denote the rate of change, namely

$$\frac{dy}{dx} = \lim_{\Delta x \to 0} \frac{\Delta y}{\Delta x} \tag{32}$$

which notationally reminds us of its origin as a limit of quotients. Moreover, he went one step further to endow meanings to the symbols dx and dy themselves so that $\frac{dy}{dx}$ becomes *itself a quotient* rather than the limit of quotients.

Definition. For an independent variable, say x, dx is defined to be Δx which is an arbitrary, small increment of x; for a dependent variable, say $y = f(x)$, dy is defined to be $f'(x) \cdot dx = f'(x) \cdot \Delta x$, namely, dy is the approximation of

$$\Delta f = f(x + \Delta x) - f(x) \tag{33}$$

up to the first order of Δx.

Remarks

(1) Conceptually, the increment of x, Δx, is just a new independent variable that varies over a domain of real numbers with small absolute values while $\Delta y = f(x + \Delta x) - f(x)$ is a function of two independent variables, namely, x and Δx. Thus $dy = f'(x) \cdot \Delta x$ is a first order approximation of Δy which is *linear* in Δx, namely, their difference $(\Delta y - dy)$ is *relatively much smaller than* $|\Delta x|$.

Examples

(i) $y = x^3$:

$$\Delta y = (x + \Delta x)^3 - x^3 = 3x^2 \Delta x + 3x \Delta x^2 + \Delta x^3$$
$$dy = 3x^2 \Delta x \; . \tag{34}$$

(ii) $y = \sin x$:

$$\Delta y = \sin(x + \Delta x) - \sin x$$
$$= \cos x \sin \Delta x + \sin x (\cos \Delta x - 1) \tag{35}$$
$$dy = \cos x \Delta x \; .$$

(iii) $y = \cos x$:

$$\Delta y = \cos(x + \Delta x) - \cos x$$
$$= -\sin x \sin \Delta x + \cos x (\cos \Delta x - 1) \tag{36}$$
$$dy = -\sin x \Delta x \; .$$

Notice that

$$\frac{|\sin \Delta x - \Delta x|}{|\Delta x|} \to 0 \text{ and } \frac{|\cos \Delta x - 1|}{|\Delta x|} \to 0 \text{ as } |\Delta x| \to 0. \tag{37}$$

(2) Let (x_0, y_0) be a point on the graph of $y = f(x)$, namely, $y_0 = f(x_0)$. Then the tangent line of the graph at (x_0, y_0) is given by the following equation, namely

$$Y - y_0 = f'(x_0)(X - x_0). \tag{38}$$

Therefore,

$$dy = f'(x)dx$$

can also be regarded as the generic representation of the collection of tangent lines of the graph of $y = f(x)$ (at its diffentiable points).

(3) In Leibnitz's system of notations, $\frac{d}{dx}$ can be used as the symbol for the operation: *differentiation* with respect to x.

In terms of such a system of notations, we summarize the basic laws of differentation and integration as follows:

(i) $\dfrac{d}{dx}(f(x) + g(x)) = \dfrac{d}{dx}f(x) + \dfrac{d}{dx}g(x)$

or $d(f(x) + g(x)) = df(x) + dg(x)$.

(ii) $\dfrac{d}{dx}(c \cdot f(x)) = c\dfrac{d}{dx}f(x)$, $c = $ constant

or $d(c \cdot f(x)) = c \cdot df(x)$.

(iii) $\dfrac{d}{dx}(f(x) \cdot g(x)) = \dfrac{d}{dx}f(x) \cdot g(x) + f(x) \cdot \dfrac{d}{dx}g(x)$

or $d(f(x) \cdot g(x)) = df(x) \cdot g(x) + f(x) \cdot dg(x)$.

(iv) $\dfrac{d}{dx}\left(\dfrac{f(x)}{g(x)}\right) = \dfrac{\dfrac{d}{dx}f(x) \cdot g(x) - f(x)\dfrac{d}{dx}g(x)}{[g(x)]^2}$

or $d\left(\dfrac{f(x)}{g(x)}\right) = \dfrac{df(x) \cdot g(x) - f(x) \cdot dg(x)}{[g(x)]^2}$.

(v) $z = g(y)$, $y = f(x)$, i.e., $z = g(f(x))$, $dz = g'(y)dy$,

$dy = f'(x)dx$, $dz = g'(y) \cdot f'(x)dx$.

Thus the chain rule of differentiation, namely

$$\frac{dz}{dx} = \frac{dz}{dy} \cdot \frac{dy}{dx}$$

becomes formally obvious.

(vi) Fundamental Theorem of Analysis:

$$d\left[\int f(x)dx\right] = f(x)dx$$

$$\int dF(x) = F(x) + C \text{ (an arbitrary constant)}.$$

(vii) $\int (f(x) + g(x))dx = \int f(x)dx + \int g(x)dx.$

(viii) $\int c \cdot f(x)dx = c \cdot \int f(x)dx.$

(ix) $\int g(f(x)) \cdot f'(x)dx = \int g(u)du, \ u = f(x).$

Exercises

If an *antiderivative* (i.e., indefinite integral) of $f(x)$ can be found, namely, a function $F(x)$ with $F'(x) = f(x)$, then it follows directly from the fundamental theorem of calculus that

$$\int_a^b f(x)dx = F(b) - F(a).$$

Using such an approach, try to compute the following definite integrals

1. $\int_a^b x^n dx = ?$

2. $\int_0^\pi \sin x dx = ?$

3. $\int_0^\pi \cos x dx = ?$

4. $\int_a^b \sin(nx)dx = ?$

5. $\displaystyle\int_a^b \cos(nx)dx =?$

   [Hint: $\dfrac{d}{dx}\left(\dfrac{1}{n}\sin(nx)\right) = \cos(nx).$]

6. $\displaystyle\int_0^{2\pi} \sin^2 x\,dx =?$

7. $\displaystyle\int_0^{2\pi} \cos^2 x\,dx =?$

   [Hint: Make use of the trigonometric identities

   $\sin^2 x = \frac{1}{2}(1 - \cos 2x)$ and $\cos^2 x = \frac{1}{2}(1 + \cos 2x).$]

8. $\displaystyle\int_0^{2\pi} \sin(mx)\cdot\sin(nx)dx =?$ (m, n: integers)

9. $\displaystyle\int_0^{2\pi} \cos(mx)\cos(nx)dx =?$ (m, n: integers)

10. $\displaystyle\int_0^{2\pi} \sin(mx)\cos(nx)dx =?$ (m, n: integers)

    [Hint: Make use of the trigonometric identities

    $$\frac{1}{2}[\cos(A + B) + \cos(A - B)] = \cos A \cos B$$
    $$\frac{1}{2}[-\cos(A + B) + \cos(A - B)] = \sin A \sin B$$
    $$\frac{1}{2}[\sin(A + B) + \sin(A - B)] = \sin A \cos B$$

    and pay attention to the special case of $m = n$ for Exercise 8 and Exercise 9.]

11. Show that

$$\frac{dx}{dy} = \frac{1}{\frac{dy}{dx}}.$$

12. Use (11) to compute

  (i) $\dfrac{d}{dy} \tan^{-1} y =?$

   [Hint: Set $x = \tan^{-1} y$; i.e., $y = \tan x$.]

  (ii) $\dfrac{d}{dy} \sin^{-1} y =?$

   [Hint: Set $x = \sin^{-1} y$; i.e., $y = \sin x$.]

CHAPTER 5

Elementary Functions and Some Typical Examples of Applications of Calculus

Calculus is itself the fundamental theory and one of the basic tools of mathematical analysis. In the quantitative analysis of systems of variable quantities, one uses functions to express the interlocking correlations amongst the variable parameters. Differentiation and integration are themselves the foundational theory and the basic operations in the study of functions. We shall conclude this concise introduction of calculus by a brief discussion on some pertinent properties of those useful elementary functions such as polynomial functions, trigonometric functions, exponential function and logarithmic function as well as some typical examples of applications of calculus. The purpose of this chapter is to provide the reader with some preliminary experience on the power of calculus and to offer some of the initial exercises on the type of computations needed in carrying out mathematical analysis.

§ 1. Elementary Functions

1.1. *Polynomial functions*

In a polynomial function $y = f(x)$, the corresponding value of

y can easily be computed from the given value of x by a simple explicit algebraic formula, namely, the given polynomial. Therefore, polynomials constitute one of the simplest and the most elementary families of function. However, they are also amongst the most useful ones. For example, it is rather easy to analyze the local properties of a polynomial function and hence it is advantageous to use the polynomial functions as the basic ones for the local approximation of general ones.

A. *Local expansion of polynomial functions and local approximation by polynomial functions*

In analyzing the local properties of a given function $f(x)$ at the point $x = a$, it is natural to make the substitution $x = a + t$, thus reducing to the convenient special location of $t = 0$. In the special case of a polynomial function, straightforward computation using the binomial formula enables us to rewrite the given polynomial $f(x)$ as another polynomial in t, namely

$$f(x) = f(a + t) = c_0 + c_1 t + c_2 t^2 + \cdots + c_n t^n. \tag{1}$$

By putting $t = 0$ on both sides, one gets $c_0 = f(a)$. If one differentiates both sides once and then puts $t = 0$, then one gets $c_1 = f'(a)$. In fact, if one differentiates both sides k times and then puts $t = 0$, then one gets the general formula that

$$c_k = \frac{1}{k!} f^{(k)}(a) \tag{2}$$

where $f^{(k)}(x)$ is the k-derivative of $f(x)$, thus obtaining the following *local expansion formula of polynomials*, namely

$$f(a + t) = f(a) + f'(a)t + \frac{f''(a)}{2!}t^2 + \cdots + \frac{f^{(k)}(a)}{k!}t^k$$

$$+ \cdots + \frac{f^{(n)}(a)}{n!}t^n \qquad (n = \deg(f)). \tag{3}$$

In the special case that $f(x) = x^n$,

$$f^{(k)}(x) = n(n-1)\ldots(n-k+1)x^{n-k} \tag{4}$$

and the above local expansion formula is exactly the well-known binomial formula. Hence, it can be regarded as a simple generalization of the binomial formula.

Notice that the absolute value of t is often restricted to be very small in "local analysis", the absolute values of higher order terms are comparatively much smaller than that of the lower order terms, thus becoming rather insignificant for most purposes. Therefore, local expansion formula of the above type is, indeed, a powerful tool in local analysis which enables us to omit those insignificant higher order terms by effectively controlling their magnitudes. Hence, it is natural to investigate whether functions of sufficiently general type also have similar local expansion formula.

A function $f(x)$, $x \in [a, b]$, is said to be of C^k-type if the k-th derivative, $f^{(k)}(x)$, exists and is continuous on $[a, b]$.

Theorem 5.1 (Taylor). *Let $f(x)$ be a function of C^k-type defined on $[a, b]$ and x_0, $x_0 + t$ be two interior points of $[a, b]$. Then there exists a suitable ξ lying between x_0 and $x_0 + t$ such that*

$$f(x_0 + t) = f(x_0) + f'(x_0)t + \frac{f''(x_0)}{2!}t^2 + \ldots$$

$$+ \frac{f^{(k-1)}(x_0)}{(k-1)!}t^{k-1} + \frac{f^{(k)}(\xi)}{k!}t^k. \tag{5}$$

Proof. In case $k = 1$, the above theorem is exactly the mean value theorem of Sec. 2, Chapter 4. Thus, the above theorem is its higher order generalization whose proof can, in fact, be reduced to a direct application of Theorem 4.4' to the following auxiliary function.

For fixed x_0 and $x_1 = x_0 + t$, set

$$R_k = f(x_1) - \left\{ f(x_0) + f'(x_0)t + \cdots + \frac{f^{(k-1)}(x_0)}{(k-1)!}t^{k-1} \right\}$$

$$F(x) = f(x_1) - f(x) - f'(x)(x_1 - x) - \frac{f''(x)}{2!}(x_1 - x)^2 - \cdots \quad (6)$$

$$- \frac{f^{(k-1)}x}{(k-1)!}(x_1 - x)^{k-1} - \frac{R_k}{t^k}(x_1 - x)^k.$$

Then, it is straightforward to check that

$$F(x_0) = 0, \ F(x_1) = 0. \quad (7)$$

Therefore, by Theorem 4.4', there exists a ξ lying between x_0 and x_1 such that $F'(\xi) = 0$. On the other hand, the differentiation of the second equation of (6) shows that

$$F'(x) = -f'(x) + [f'(x) - f''(x)(x_1 - x)]$$

$$+ \left[f''(x)(x_1 - x) - \frac{f^3(x)}{2!}(x_1 - x)^2 \right] + \cdots$$

$$+ \left[\frac{f^{(k-1)}(x)}{(k-2)!}(x_1 - x)^{k-2} - \frac{f^{(k)}(x)}{(k-1)!}(x_1 - x)^{k-1} \right]$$

$$+ \frac{kR_k(x_1 - x)^{k-1}}{t^k}. \quad (8)$$

Therefore, it follows from $F'(\xi) = 0$ that

$$-\frac{f^{(k)}(\xi)}{(k-1)!} + k \cdot \frac{R_k}{t^k} = 0 , \quad (9)$$

namely

$$R_k = \frac{f^{(k)}(\xi)}{k!}t^k. \quad (9')$$

$\square$

Remarks

(i) The above important theorem of Taylor collectively and concisely reveals the wholesome meaning of the successive derivatives of a given function at a given point. A function of C^k-type can be locally approximated up to the k-th order by a suitable polynomial function of degree at most k, namely

$$f(a+t) = f(a) + f'(a)t + \frac{f''(a)}{2!}t^2 + \cdots + \frac{f^{(k)}(a)}{k!}t^k. \qquad (10)$$

(ii) If $R_k \to 0$ as $k \to \infty$ for a specific function $f(x)$ over a suitable range of x, then the limiting case of the above local expansion is called a power series expansion of such a function, namely

$$f(x) = f(a) + f'(a)\cdot(x-a) + \frac{f''(a)}{2!}(x-a)^2 + \cdots + \frac{f^{(k)}(a)4}{k!}(x-a)^k + \cdots \qquad (11)$$

(The Taylor's series of $f(x)$ at $x = a$, the special case of $a = 0$ is often called the Maclaurin series of $f(x)$.)

B. *Continuity and roots of polynomial functions*

Let $f(x)$ be a polynomial of degree n with real coefficients and $y = f(x)$ be the polynomial function of a real variable $x \in \mathbb{R}$. Then the real roots of $f(x)$ are exactly the x-coordinates of the intersection points of the graph of $y = f(x)$ and the x-axis. By the continuity of $f(x)$ and the intermediate value theorem of continuous function, $f(x)$ has at least one real root between a and b if $f(a) \cdot f(b) < 0$ (i.e., $f(a)$ and $f(b)$ are of opposite signs). Now, we shall seek a deeper result of determining the exact number of real roots of $f(x)$ lying between two given numbers a and b. Recall that α is called a multiple root of $f(x)$ if $f(x)$ is divisible by $(x-\alpha)^2$, namely, $f(x) = (x-\alpha)^2 \cdot g(x)$. In such a case, one has

$$f'(x) = 2(x-\alpha)g(x) + (x-\alpha)^2 g'(x)$$

$$= (x-\alpha)[2g(x) + (x-\alpha)\cdot g'(x)] \qquad (12)$$

and hence $(x - \alpha)$ is a common divisor of $f(x)$ and $f'(x)$. Therefore, $f(x)$ contains no multiple roots if $f(x)$ and $f'(x)$ are relatively prime. In case $f(x)$ and $f'(x)$ are *not* relatively prime, set $d(x)$ to be the greatest common divisor of $f(x)$ and $f'(x)$ and $f_0(x) = f(x)/d(x)$. Then it is not difficult to show that $f(x)$ and $f_0(x)$ have the *same set of roots* but $f_0(x)$ and $f_0'(x)$ are relatively prime. Therefore, we may assume without loss of generality that $f(x)$ and $f'(x)$ are relatively prime to begin with.

Definition. To a given real coefficient polynomial $f(x)$ with $(f(x), f'(x)) = 1$ (i.e., relatively prime), set

$$f_0(x) = f(x), \quad f_1(x) = f'(x) \tag{13}$$

and inductively by division algorithm such that

$$f_{k-1}(x) = q_k(x)f_k(x) - f_{k+1}(x), \quad \deg(f_{k+1}) < \deg(f_k). \tag{13'}$$

By the assumption $(f(x), f'(x)) = 1$, the last one $f_l(x)$ is a non-zero constant. The above sequence of polynomials $\{f_0(x), f_1(x), \ldots, f_l(x)\}$ is called the Sturm sequence of $f(x)$.

Theorem 5.2 (Sturm). *The number of real roots of the above $f(x)$ lying between a and b is equal to $V(a) - V(b)$ where $V(a)$ (resp. $V(b)$) is the number of sign changes in the following sequence of numbers, namely*

$$\{f_0(a), f_1(a), \ldots, f_l(a)\} \ (resp. \ \{f_0(b), f_1(b), \ldots, f_l(b)\}). \tag{14}$$

Proof. Notice that the sequence of numbers

$$\{f_0(\xi), f_1(\xi), \ldots, f_l(\xi)\} \tag{15}$$

cannot have consecutive zeros, because $f_{k-1}(\xi) = f_k(\xi) = 0$ and (13′) implies that $f_{k+1}(\xi) = 0$, thus implying that $f_l(\xi) = 0$ (which is impossible because $f_l(x)$ is itself a non-zero constant). Moreover, $f_k(\xi) = 0$ and (13′) implies that $f_{k-1}(\xi)$ and $f_{k+1}(\xi)$ are numbers of opposite signs. Therefore the number of sign changes, $V(\xi)$, in the sequence $\{f_i(\xi); \ 0 \le i \le l\}$ is always well-defined even if the above sequence contains some zeros.

Suppose c is a root of $f_k(x)$, $k > 0$. Then $f_{k-1}(c) = -f_{k+1}(c) \neq 0$. Therefore, for a sufficiently small δ, one has, by continuity, the following sign-distribution for the segment $\{f_{k-1}, f_k, f_{k+1}\}$, namely

x	$c - \delta$	c	$c + \delta$
f_{k-1}	$+$	$+$	$+$
f_k	$\pm$	0	$\pm$
f_{k+1}	$-$	$-$	$-$

or

x	$c - \delta$	c	$c + \delta$
f_{k-1}	$-$	$-$	$-$
f_k	$\pm$	0	$\pm$
f_{k+1}	$+$	$+$	$+$

Thus, no matter what the signs of $f_k(c \pm \delta)$ may be, the contribution of sign changes of the segment $\{f_{k-1}, f_k, f_{k+1}\}$ toward $V(c - \delta)$ and $V(c + \delta)$ are always equal to 1.

Next let us consider the only remaining case that c is a root of $f_0(x)$. Then $f_1(c)$ must be *non-zero*. Recall that $f_1(x)$ is, by definition, the derivative of $f_0(x)$. Therefore $f_0(x)$ is monotonically increasing (resp. decreasing) in a small neighborhood of c if $f_1(c)$ is positive (resp. negative). Hence, the sign distribution of the beginning segment $\{f_0, f_1\}$ has the following two possibilities, namely

x	$c - \delta$	c	$c + \delta$
f_0	$-$	0	$+$
f_1	$+$	$+$	$+$

or

$$\begin{array}{c|ccc} x & c-\delta & c & c+\delta \\ \hline f_0 & + & 0 & - \\ f_1 & - & - & - \end{array}$$

Anyway, the contribution of sign change of the segment $\{f_0, f_1\}$ is equal to 1 toward $V(c-\delta)$ but becomes 0 toward $V(c+\delta)$.

Now, let us consider the integral valued function $V(\xi)$ which records the number of sign changes of the sequence $\{f_i(\xi); 0 \le i \le l\}$ for $a \le \xi \le b$. The above discussion proves that $V(\xi)$ remains to be constant except at the vicinity of a root of $f(x)$ itself, and moreover

$$V(c-\delta) = V(c+\delta) + 1. \tag{16}$$

Hence $V(a) - V(b)$ is exactly the number of real roots of lying between a and b.

Examples

1. Suppose that a degree n polynomial of real coefficients $f(x)$ has exactly n distinct real roots. Then, by Theorem 5.2, $V(-K) = n$ and $V(K) = 0$ for all sufficiently large K. First of all, the Sturm sequence of $f(x)$ must consist of $(n+1)$ polynomials, i.e., $l = n$, and hence $\deg(f_k(x)) = n - k$. Secondly, the leading coefficients of $f_k(x)$, $0 \le k \le n$, must be all of the same sign. For otherwise, $V(K) > 0$ for sufficiently large K.

[Exercise: Show that the above two conditions are also sufficient for $f(x)$ to have n distinct real roots, $n = \deg(f)$.]

2. Suppose that a degree n polynomial $f(x)$ has exactly n distinct positive real roots. Then, by Theorem 5.2, $V(0) = n$ and $V(K) = 0$ for all sufficiently large K.

[Exercise: What should be the necessary and sufficient condition on the coefficients of the Sturm sequence of $f(x)$ for $f(x)$ to have n positive distinct real roots?]

3. $f_0(x) = x^3 - x^2 - 10x + 1$. Then

$$f_1(x) = 3x^2 - 2x - 10, \quad f_2(x) = 62x + 1, \quad f_3(x) = 38313$$

x	$-K$	-3	-2	0	2	4	$+K$
f_0	$-$	$-$	$+$	$+$	$-$	$+$	$+$
f_1	$+$	$+$	$+$	$-$	$-$	$+$	$+$
f_2	$-$	$-$	$-$	$+$	$+$	$+$	$+$
f_3	$+$	$+$	$+$	$+$	$+$	$+$	$+$

Therefore $f_0(x)$ has three real roots and they are located in the intervals $[-3, -2]$, $[0, 2]$ and $[2, 4]$ respectively.

4. $f_0(x) = x^5 - 5x^4 + 9x^3 - 9x^2 + 5x - 1$. Then

$$f_1(x) = 5x^4 - 20x^3 + 27x^2 - 18x + 5$$
$$f_2(x) = x^3 - x$$
$$f_3(x) = -32x^2 + 38x - 5$$
$$f_4(x) = -26x + 19$$
$$f_5(x) = -192.$$

Therefore, it is not difficult to check that

$$V(-K) = 4, \ V(K) = 1$$

for sufficiently large K. Hence $f_0(x)$ has three real roots.

[Exercise: Try to locate the positions of its three real roots up to intervals of consecutive integers.]

It is well known that a quadratic polynomial such as $x^2 - 2ax + (a^2 + b^2)$ has no real roots. However, any quadratic polynomial always has complex roots such as $a \pm b_i$ for the above one. How about polynomials of higher degrees, do they also always have complex roots? Such a question was asked and believed to be the case since the time of Euler, and was finally proved to be the case by Gauss—the fundamental theorem of algebra. As an application of continuity, we present here one of its elementary proofs.

The Fundamental Theorem of Algebra (Gauss). *To any given complex (or real) coefficient polynomial of positive degree $f(z)$, there always exists at least one complex root z_0, i.e., $f(z_0) = 0$.*

Proof. Let

$$f(z) = z^n + a_1 z^{n-1} + \cdots + a_i z^{n-i} + \cdots + a_n, \ a_i \in \mathbb{C}. \qquad (17)$$

We shall assume that $n > 1$ and $a_n \neq 0$. Otherwise, the proof is trivial. Set

$$M = \text{Max}\{|a_i|; \ 1 \le i \le n\}$$
$$D = \{z = x + iy; \ |x|, |y| \le 2M + 1\}. \qquad (18)$$

D is a square of size $(4M+2) \times (4M+2)$ in the complex plane centered at the origin. We shall show that there exists an interior point z_0 of D with $f(z_0) = 0$. Consider $f(x + iy)$ as a complex-value function of two real variables x and y and set

$$f(x + iy) = u(x, y) + iv(x, y)$$
$$|f(x + iy)| = \{u(x, y)^2 + v(x, y)^2\}^{1/2} = w(x, y) \qquad (19)$$

where $u(x, y)$ and $v(x, y)$ are real coefficient polynomials in x and y. Set

$$L(D) = \text{g.l.b.}\{w(x, y); \ (x, y) \in D\} \ (\le w(0, 0) = |f(0)| = |a_n| \le M). \qquad (20)$$

Then, the proof consists of the following three crucial steps, namely

(i) There exists a minimal point of $w(x, y)$, namely $(x_0, y_0) \in D$ such that $w(x_0, y_0) = L(D)$.

(ii) The values of $w(x, y)$ on the boundary points of D are always larger than M and hence the above minimal point (x_0, y_0) must be an interior point.

(iii) Using the fact $w(x, y) = |f(z)|$ to show that $w(x_0, y_0) = |f(x_0 + iy_0)| = 0$.

The Proof of (i). The proof here is essentially the same as that of Theorem 4.3. The only modification needed is that of cutting a square D_n into four quarter-subsquares instead of cutting an interval I_n into two half-subintervals. Again, the *guiding principle* of the inductive choice of D_{n+1} out of the four subsquares of D_n is to preserve the property that

$$L(D_{n+1}) = L(D_n) = \cdots = L(D), \tag{21}$$

thus obtaining a nest of squares

$$D \supset D_2 \supset D_3 \supset \cdots \supset D_n \supset D_{n+1} \supset \cdots$$

In terms of coordinates, one has two pairs of approaching sequences of real numbers $\{a_n\}$, $\{b_n\}$ and $\{c_n\}$, $\{d_n\}$ such that

$$\begin{cases} a_1 \le a_2 \le \cdots \le a_n \le \cdots \le b_n \le \cdots \le b_2 \le b_1 \\ c_1 \le c_2 \le \cdots \le c_n \le \cdots \le d_n \le \cdots \le d_2 \le d_1 \\ (b_n - a_n) = (d_n - c_n) = \left(\tfrac{1}{2}\right)^{n-1} \cdot (4M + 2) \to 0 \\ D_n = \{(x,y);\ a_n \le x \le b_n,\ c_n \le y \le d_n\}. \end{cases} \tag{22}$$

Again, by the completeness of the real number system, there exists $x_0, y_0 \in \mathbb{R}$ such that

$$a_n \to x_0 \leftarrow b_n \text{ and } c_n \to y_0 \leftarrow d_n. \tag{23}$$

Finally, it follows from (21) and the continuity of $w(x,y)$ that $w(x_0, y_0) = L(D)$.

Proof of (ii). Suppose that $(x,y) \in \partial D$ is a boundary point of D. Then $|z| = \sqrt{x^2 + y^2} \ge (2M + 1)$. Therefore,

$$w(x,y) = |f(z)| \ge |z^n| - |a_1 z^{n-1} + \cdots + a_n|$$

$$\ge |z|^n - \frac{M}{|z| - 1}(|z|^n - 1),\ |z| \ge 2M + 1 \tag{24}$$

$$\ge |z|^n - \frac{1}{2}(|z|^n - 1) \ge \frac{1}{2}|z|^n > M(2M + 1)^{n-1} > M.$$

Hence (x_0, y_0) cannot be a boundary point of D, namely, (x_0, y_0) must be an interior point.

Proof of (iii). Finally, we shall prove that $w(x_0, y_0) = |f(z_0)| = L$ must be zero by deducing a contradiction from the assumption that $L \neq 0$. Set

$$z_1 = z_0 + r(\cos\theta + i\sin\theta) = z_0 + t$$

where r is a sufficiently small number so that z_1 is in the interior of D for all θ. Set

$$
\begin{aligned}
f(z_1) = f(z_0 + t) &= f(z_0) + c_1 t + c_2 t^2 + \cdots + t^n \\
&= f(z_0) + c_k t^k \cdot [1 + c'_{k+1} t + \cdots + c'_n t^{n-k}] \quad (25) \\
&= f(z_0) + c_k t^k \cdot g(t)
\end{aligned}
$$

where c_k is the first non-zero c_i and $c'_i = c_i / c_k$. Set

$$K = \text{Max}\{|c'_i|, \; k+1 \leq i \leq n\} \quad (26)$$

and choose r to be smaller than $1/(2K+1)$. Thus

$$\left(K + \frac{1}{2}\right) \cdot r < \frac{1}{2} \Rightarrow Kr < \frac{1}{2}(1-r) \Rightarrow K \cdot \frac{r}{1-r} < \frac{1}{2} \quad (27)$$

and hence

$$|c'_{k+1} t + \cdots + c'_n t^{n-k}| \leq K \cdot (r + \cdots + r^{n-k}) < Kr \cdot \frac{1}{1-r} < \frac{1}{2}. \quad (28)$$

Now, fix r and let θ vary over $[0, 2\pi]$, namely, let $z_1 = z_0 + r(\cos\theta + i\sin\theta)$ vary over a small circle in the interior of D. Set A, B and $\Lambda(t)$ to be the absolute values of $f(z_0)$, c_k and $g(t)$ and $\alpha, \beta, \lambda(t)$ to be the amplitudes of $f(z_0)$, c_k and $g(t)$ respectively, namely

$$
\begin{cases}
f(z_0) = A \cdot (\cos\alpha + i\sin\alpha) \\
c_k = B \cdot (\cos\beta + i\sin\beta) \\
g(t) = 1 + c'_{k+1} t + \cdots + c'_n t^{n-k} = \Lambda(t) \cdot (\cos\lambda(t) + i\sin\lambda(t)).
\end{cases}
\quad (29)
$$

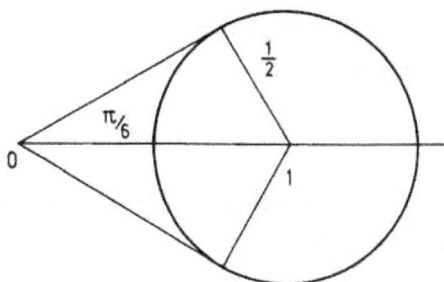

Fig. 33

It follows from (28) that $g(t)$ always lies inside of the circle of radius $\frac{1}{2}$ centered at 1. Therefore, as indicated in Fig. 33, $|\lambda(t)| < \pi/6$.

Notice that

$$
\begin{aligned}
f(z_1) - f(z_0) &= B(\cos\beta + i\sin\beta) \cdot r^k \cdot (\cos k\theta + i\sin k\theta) \\
&\quad \cdot \Lambda(t) \cdot (\cos\lambda(t) + i\sin\lambda(t)) \\
&= B \cdot r^k \cdot \Lambda(t) \\
&\quad \cdot [\cos(\beta + k\theta + \lambda(t)) + i\sin(\beta + k\theta + \lambda(t))].
\end{aligned}
\tag{30}
$$

Now, choose $\theta = \frac{1}{k}(\pi + \alpha - \beta)$. Then

$$
|\beta + k\theta + \lambda(t) - (\pi + \alpha)| = |\lambda(t)| < \pi/6 \tag{31}
$$

namely

$$
\pi + \alpha - \pi/6 < \beta + k\theta + \lambda(t) < \pi + \alpha + \pi/6. \tag{31'}
$$

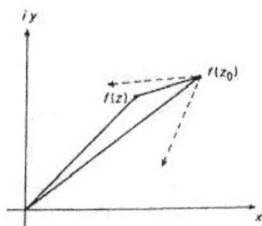

Fig. 34

As indicated in Fig. 34, the vector $f(z_1) - f(z_0)$ is a directed interval confined by the two dotted lines. Therefore, it is obvious that

$$|f(z_1)| < |f(z_0)| = L(D)$$

which is a contradiction to the assumption that $L(D)$ is the g.l.b. of $|f(z)|$ on D. This proves that $f(z_0) = L = 0$. □

Corollary 1. *Every positive degree complex coefficient polynomial $f(z) \in \mathbb{C}[z]$ can be factored into the product of linear polynomials with complex coefficients.*

Corollary 2. *Every positive degree real coefficient polynomial $f(x) \in \mathbb{R}[x]$ can be factored into the product of linear or quadratic polynomials with real coefficients.*

[Exercises: Prove Corollary 1 and 2.]

1.2. Trigonometric functions

Let t be the time parameter and x, y be the Cartesian coordinates. Then the motion of a point on the unit circle with unit speed can be described by two basic functions, namely

$$x = \cos t, \quad y = \sin t \tag{32}$$

The above circular motion with uniform speed is the simplest and the most basic periodic motion and the pair of functions describing such a motion, namely, the sine and the cosine, are the most basic *periodic functions*. The basic properties of sine and cosine are closely related to the geometric properties of the unit circle. For examples,

(i) The equation of unit circle $x^2 + y^2 = 1$ corresponds to the functional identity:

$$\cos^2 t + \sin^2 t \equiv 1. \tag{33}$$

(ii) The basic identity

$$\cos(\alpha - \beta) = \cos \alpha \cos \beta + \sin \alpha \sin \beta \tag{34}$$

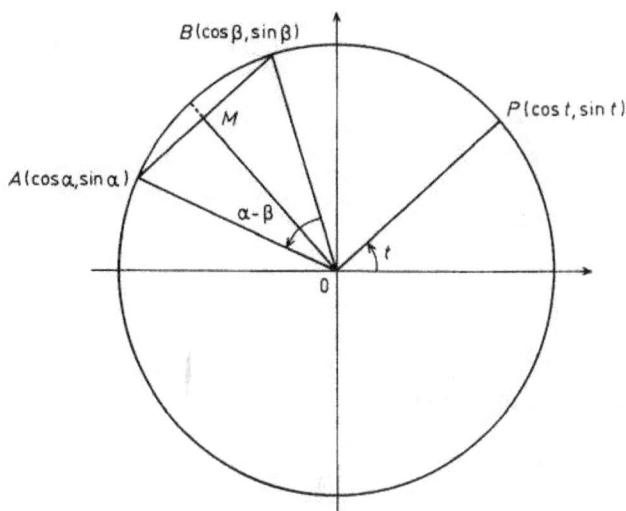

Fig. 35

is a simple consequence of the rotational symmetry of the unit circle.

Proof. It follows from the rotational symmetry that the distance between $A(\cos\alpha, \sin\alpha)$ and $B(\cos\beta, \sin\beta)$ *only depends* on the difference of angles. Therefore

$$(\cos\alpha - \cos\beta)^2 + (\sin\alpha - \sin\beta)^2 = (\cos(\alpha - \beta) - 1)^2 + \sin^2(\alpha - \beta) \quad (35)$$

namely

$$2 - 2(\cos\alpha\cos\beta + \sin\alpha\sin\beta) = 2 - 2\cos(\alpha - \beta). \quad (36)$$

$\square$

(iii) The velocity vector of the above unit speed circular motion is clearly the unit tangent vector at $P(\cos t, \sin t)$. It is easy to see that the two components of such a vector are equal to

$$\cos\left(t + \frac{\pi}{2}\right) = -\sin t \quad \text{and} \quad \sin\left(t + \frac{\pi}{2}\right) = \cos t. \quad (37)$$

On the other hand, the components of the velocity vector should be equal to the derivatives of $\cos t$ and $\sin t$ respectively. Hence

$$\frac{d}{dt}\cos t = -\sin t, \quad \frac{d}{dt}\sin t = \cos t. \tag{38}$$

(iv) Let M be the middle point of $\overline{AB}$. Then the

$$\overrightarrow{OM} = \left(\frac{1}{2}(\cos\alpha + \cos\beta), \frac{1}{2}(\sin\alpha + \sin\beta)\right). \tag{39}$$

On the other hand, it is well-known that $\overrightarrow{OM}$ bisects $\angle AOB$. Hence

$$\begin{aligned}
\overrightarrow{OM} &= |\overrightarrow{OM}| \cdot \left(\cos\frac{1}{2}(\alpha+\beta), \sin\frac{1}{2}(\alpha+\beta)\right)\\
&= \cos\frac{1}{2}(\alpha-\beta) \cdot \left(\cos\frac{1}{2}(\alpha+\beta), \sin\frac{1}{2}(\alpha+\beta)\right)
\end{aligned} \tag{39'}$$

thus obtaining the formula

$$\begin{cases}
\cos\alpha + \cos\beta = 2\cos\frac{1}{2}(\alpha-\beta)\cos\frac{1}{2}(\alpha+\beta)\\
\sin\alpha + \sin\beta = 2\cos\frac{1}{2}(\alpha-\beta)\sin\frac{1}{2}(\alpha+\beta).
\end{cases} \tag{40}$$

Using (iii), it is straightforward to compute the Maclaurin series of $\sin x$ and $\cos x$, namely

Theorem 5.3.

$$\begin{aligned}
\sin x &= x - \frac{x^3}{3!} + \frac{x^5}{5!} - \cdots + (-1)^n\frac{x^{2n+1}}{(2n+1)!} + \cdots\\
\cos x &= 1 - \frac{x^2}{2!} + \frac{x^4}{4!} - \cdots + (-1)^n\frac{x^{2n}}{(2n)!} + \cdots
\end{aligned} \tag{41}$$

Proof. Set $f(x) = \sin x$ (resp. $\cos x$). Then

$$f(0) = 0, \; f'(0) = 1, \; f''(0) = 0, \; f^{(3)}(0) = -1$$

(resp. $f(0) = 1$, $f'(0) = 0$, $f''(0) = -1$, $f^{(3)}(0) = 0$) and $f^{(k+4)}(x) = f^{(k)}(x)$. $\square$

1.3. *Exponential function and logarithmic function*

Let us begin with a brief discussion on *compounded interest*. Suppose the interest rate is x for a given period, say 10 years. If it is computed just as a simple interest, then the amount of money one will get at the end of the period is simply equal to $(1+x)$ times the amount of deposit. However, if it is computed as a compounded-yearly interest, then the amount one will get at the end of the 10-year period will be equal to $\left(1 + \frac{x}{10}\right)^{10}$ times the amount of deposit. Similarly, if it is computed as a compounded-monthly interest, then the factor will be equal to

$$\left(1 + \frac{x}{120}\right)^{120}.$$

In general, if the whole period is divided into n subunit and one computes the compounded interest accordingly, then the factor will be equal to

$$\left(1 + \frac{x}{n}\right)^n, \tag{42}$$

which is a polynomial function of degree n, $f_n(x)$. Using binomial formula, the coefficients of the above polynomial $f_n(x)$ can be computed as follows

$$\begin{aligned}
f_n(x) &= \left(1 + \frac{x}{n}\right)^n = 1 + n \cdot \frac{x}{n} + \frac{n(n-1)}{2!}\left(\frac{x}{n}\right)^2 + \cdots \\
&= 1 + x + 1 \cdot \left(1 - \frac{1}{n}\right)\frac{x^2}{2!} + \left(1 - \frac{1}{n}\right)\left(1 - \frac{2}{n}\right)\frac{x^3}{3!} + \cdots \\
&\quad + \left(1 - \frac{1}{n}\right)\left(1 - \frac{2}{n}\right)\cdots\left(1 - \frac{k-1}{n}\right)\frac{x^k}{k!} + \cdots
\end{aligned} \tag{43}$$

Therefore, as $n \to \infty$, the coefficient of degree k in $f_n(x)$ tends to $\frac{1}{k!}$ as a limit. Hence, the limiting function, namely

$$f(x) = \lim_{n \to \infty} f_n(x) \tag{44}$$

can be represented by the following power series, namely

$$f(x) = 1 + x + \frac{x^2}{2!} + \frac{x^3}{3!} + \cdots + \frac{x^k}{k!} + \cdots \tag{44'}$$

Intuitively speaking, $f(x)$ is the factor if the interest is compounded *continuously*! This function has remarkable properties and plays an important role in mathematics.

(a) $f(mx) = [f(x)]^m$:

$$f_{mn}(mx) = \left(1 + \frac{mx}{mn}\right)^{mn} = \left[\left(1 + \frac{x}{n}\right)^n\right]^m = [f_n(x)]^m.$$

Hence, as $n \to \infty$, one gets

$$f(mx) = [f(x)]^m$$

(b) $f\left(\frac{m}{l} \cdot x\right) = [f(x)]^{\frac{m}{l}}$:

$$\left[f\left(\frac{m}{l}x\right)\right]^l = f(mx) = [f(x)]^m \text{ (by (1))}.$$

Hence

$$f\left(\frac{m}{l}x\right) = [f(x)]^{\frac{m}{l}}.$$

Theorem 5.4. *The function $f(x)$ of (44') is equal to e^x where*

$$e = f(1) = 1 + 1 + \frac{1}{2!} + \frac{1}{3!} + \cdots + \frac{1}{k!} + \cdots \tag{45}$$

and it has the following basic properties, namely

(i) $f(x)$ *is monotonically increasing*

(ii) $f(x_1 + x_2) = f(x_1) \cdot f(x_2)$

(iii) $f'(x) = f(x)$.

Proof. It is easy to check that $f(x)$ is monotonically increasing and $\lim_{x \to 0} f(x) = 1$.

Set $e = f(1) = 2.71828182\ldots$. By the above (b), one has

$$f\left(\frac{m}{l}\right) = e^{\frac{m}{l}}. \tag{46}$$

Hence, it follows from the *monotonicity* of *both* $f(x)$ *and* e^x that

$$f(x) = e^x \tag{46'}$$

for all $x \in \mathbb{R}$.

Therefore, it is simply the exponent rule that

$$f(x_1 + x_2) = e^{x_1 + x_2} = e^{x_1} \cdot e^{x_2} = f(x_1) \cdot f(x_2).$$

Finally, on the one hand, it follows directly from (44') that $f'(0) = 1$ and on the other hand

$$f'(x_0) = \lim_{h \to 0} \frac{f(x_0 + h) - f(x_0)}{h}$$

$$\lim_{h \to 0} \frac{f(x_0)(f(h) - 1)}{h} = f(x_0) \lim_{h \to 0} \frac{f(h) - 1}{h} \tag{47}$$

$$= f(x_0) \cdot f'(0) = f(x_0).$$

$\square$

Remarks

(i) If one multiplies the power series of $f(x)$ and $f(y)$ by the usual rule of algebra, namely

$$f(x) \cdot f(y) = \left(\sum_{i=0}^{\infty} \frac{x^i}{i!} \right) \left(\sum_{j=0}^{\infty} \frac{y^j}{j!} \right)$$

$$= \sum_{k=0}^{\infty} \sum_{i+j=k} \frac{x^i y^j}{i! j!} \tag{48}$$

and then compares it with

$$f(x + y) = \sum_{k=0}^{\infty} \frac{(x + y)^k}{k!} \tag{48'}$$

what one gets is the binomial formula for each degree k.

(ii) $f(x) = e^x$ is a function whose derivative is just itself! What are those functions whose derivatives are just themselves?

Notation. Let D be the differentiation operator which maps a (smooth) function $f(x)$ to its derivative $f'(x)$ and D^k is the usual composition of applying D k times, namely

$$Df(x) = f'(x)$$
$$D^k f(x) = D(D^{k-1}f(x)) = f^{(k)}(x). \tag{49}$$

Moreover, to each polynomial of D

$$p(D) = a_0 D^n + a_1 D^{n-1} + \cdots + a_n \tag{50}$$

one has the differential operator given by

$$p(D) \cdot f(x) = a_0 \cdot D^n f(x) + a_1 \cdot D^{n-1} f(x) + \cdots + a_n \cdot f(x). \tag{50'}$$

Definition. $f(x)$ is called a solution of the differential equation

$$p(D) \cdot y = 0 \tag{51}$$

if $p(D) \cdot f(x)$ is the zero function.

Theorem 5.5. *$f(x)$ is a solution of the differential equation*

$$(D - k)^n y = 0, \quad (k \text{ a real constant}) \tag{52}$$

if and only if

$$f(x) = e^{kx} \cdot g(x) \tag{53}$$

where $g(x)$ is a polynomial of degree at most $(n - 1)$.

Proof. Let $f(x)$ be a solution of (52) and set

$$g(x) = e^{-kx} \cdot f(x), \quad (\text{i.e.}, \ f(x) = e^{kx} \cdot g(x)). \tag{54}$$

Then

$$(D - k) \cdot f(x) = D(e^{kx} \cdot g(x)) - ke^{kx}g(x)$$
$$= ke^{kx}g(x) + e^{kx}Dg(x) - ke^{kx}g(x) = e^{kx}Dg(x) \tag{55}$$

and hence, inductively

$$(D - k)^n f(x) = (D - k)[(D - k)^{n-1}f(x)]$$
$$= (D - k)[e^{kx} \cdot D^{n-1}g(x)] = e^{kx}D^n g(x). \tag{56}$$

Therefore,

$$(D - k)^n f(x) \equiv 0 \text{ if and only if } D^n g(x) \equiv 0.$$

It follows from Theorem 5.1 that $D^n g(x) = 0$ if and only if $g(x)$ is a polynomial of degree at most $(n - 1)$. $\square$

Theorem 5.6. *Let $p_1(X)$ and $p_2(X)$ be two relatively prime polynomials. Then every solution $f(x)$ of*

$$p_1(D) \cdot p_2(D)y = 0 \tag{57}$$

can be uniquely expressed as the sum of solutions of

$$p_1(D)y = 0 \text{ and } p_2(D)y = 0. \tag{58}$$

Proof. Let $f(x)$ be a given function satisfying

$$(p_1(D) \cdot p_2(D))f(x) \equiv 0. \tag{59}$$

We need to show that there exists a *unique* pair of functions $f_1(x)$ and $f_2(x)$ such that

$$p_1(D)f_1(x) \equiv 0, \ p_2(D)f(x) \equiv 0 \text{ and } f(x) = f_1(x) + f_2(x). \tag{60}$$

By the assumption that the G.C.D. of $p_1(X)$ and $p_2(X)$ is 1 and the Euclid Algorithm, there exists a pair of polynomials $u_1(X)$ and $u_2(X)$ such that

$$1 \equiv u_1(X) \cdot p_1(X) + u_2(X) \cdot p_2(X). \tag{61}$$

Therefore,

$$1 = u_1(D) \cdot p_1(D) + u_2(D)p_2(D) \tag{61'}$$

and hence

$$f(x) = (u_1(D)p_1(D))f(x) + (u_2(D)p_2(D))f(x). \tag{62}$$

Set

$$f_1(x) = (u_2(D)p_2(D))f(x), \ \ f_2(x) = (u_1(D)p_1(D))f(x). \tag{63}$$

Then

$$\begin{aligned}
p_1(D)f_1(x) &= (p_1(D)u_2(D)p_2(D))f(x) \\
&= u_2(D)(p_1(D)p_2(D))f(x) \equiv 0 \\
p_2(D)f_2(x) &= p_2(D)(u_1(D)p_1(D))f(x) \\
&= u_1(D)(p_1(D)p_2(D))f(x) \equiv 0.
\end{aligned} \tag{64}$$

This proves the existence of $\{f_1(x), f_2(x)\}$ satisfying (61). Next let us prove the uniqueness of such a pair of functions. Suppose that $\{\hat{f}_1(x), \hat{f}_2(x)\}$ is another pair of functions satisfying (60). Then

$$g(x) = f_1(x) - \hat{f}_1(x) = \hat{f}_2(x) - f_2(x) \tag{65}$$

is a function satisfying

$$p_1(D)g(x) \equiv 0 \text{ and } p_2(D)g(x) \equiv 0. \tag{66}$$

Again, by using the algebraic identity (61)

$$g(x) = 1 \cdot g(x) = u_1(D)p_1(D)g(x) + u_2(D)p_2(D)g(x) \equiv 0 \tag{67}$$

namely, $g(x)$ is the zero function and hence

$$f_1(x) = \hat{f}_1(x) \text{ and } f_2(x) = \hat{f}_2(x).$$

This proves the uniqueness of the decomposition. □

In view of the above theorem and the fundamental theorem of algebra, it is natural to seek generalization of Theorem 5.5 for *complex* constant k, namely, what are the solutions of the differential equation

$$(D - k)^n y = 0$$

where $k = a + bi$ is a complex constant? Of course, such a function must be a *complex-value* function of a real variable x, namely, $f(x) = g(x) + i \cdot h(x)$ where $g(x)$ and $i \cdot h(x)$ are the real and imaginary part of $f(x)$. For example

$$\begin{aligned}
f_1(x) &= \cos x + i \sin x \\
f_2(x) &= e^{ax} \cos bx + i e^{ax} \sin bx
\end{aligned} \tag{68}$$

are this type of function. Moreover

$$\begin{aligned}
Df_1(x) &= D \cos x + i D \sin x \\
&= -\sin x + i \cos x = i f_1(x) \\
Df_2(x) &= (De^{ax})(\cos bx + i \sin bx) + e^{ax} \cdot D(\cos bx + i \sin bx) \quad (69) \\
&= a \cdot e^{ax}(\cos bx + i \sin bx) + bi \cdot e^{ax}(\cos bx + i \sin bx) \\
&= (a + bi)f_2(x).
\end{aligned}$$

Theorem 5.5′. $f(x)$ *is a solution of the differential equation*

$$(D - (a + b_i))^n y = 0 \tag{70}$$

if and only if

$$f(x) = e^{ax}(\cos bx + i \sin bx) \cdot g(x) \tag{71}$$

where $g(x)$ is a polynomial of degree at most $(n-1)$.

Proof. Again, set

$$g(x) = e^{-ax}(\cos bx - i \sin bx)f(x)$$
$$(\text{or } f(x) = e^{ax}(\cos bx + i \sin bx)g(x)). \tag{72}$$

Then

$$(D - (a+bi))f(x) = e^{ax}(\cos bx + i \sin bx)Dg(x) \tag{73}$$

and hence, inductively

$$[D - (a+bi)]^n f(x) = e^{ax}(\cos bx + i \sin bx)D^n g(x). \tag{73'}$$

Therefore $f(x)$ is a solution of (70) if and only if $g(x)$ is a solution of $D^n y = 0$, namely, $g(x)$ is a polynomial of degree at most $(n-1)$.

$\square$

Definition. $e^{(a+ib)x}$ is defined to be $e^{ax}(\cos bx + i \sin bx)$. In particular $e^{ix} = \cos x + i \sin x$, $e^{-ix} = \cos x - i \sin x$, $e^{i\pi} = -1$.
Euler's Formula

$$\cos x = \frac{1}{2}(e^{ix} + e^{-ix})$$
$$\sin x = \frac{1}{2i}(e^{ix} - e^{-ix}). \tag{74}$$

The combined addition formula in sine and cosine, namely

$$\cos(x+y) + i \sin(x+y) = (\cos x + i \sin x)(\cos y + i \sin y)$$

now becomes the familiar form of exponent rule, i.e.,

$$e^{i(x+y)} = e^{ix} \cdot e^{iy}. \tag{75}$$

Finally, in concluding this section of brief introduction of elementary functions, we introduce the natural logarithmic function which is, by definition, the inverse function of the exponential function e^x.

Definition. $x = \ln y \Leftrightarrow y = e^x$.

Theorem 5.7. *The natural logarithmic function* $\ln y$, $y > 0$, *satisfies the following basic properties*
 (i) $\ln(y_1 \cdot y_2) = \ln y_1 + \ln y_2$
 (ii) *monotonically increasing*
(iii) $\frac{d}{dy} \ln y = \frac{1}{y}$.

Proof. Set

$$x_1 = \ln y_1, \quad x_2 = \ln y_2.$$

Then

$$y_1 = e^{x_1}, \quad y_2 = e^{x_2}, \quad y_1 \cdot y_2 = e^{x_1} \cdot e^{x_2} = e^{x_1 + x_2}$$

and hence

$$\ln y_1 y_2 = x_1 + x_2 = \ln y_1 + \ln y_2. \tag{76}$$

Notice that $e^x > 1$ if and only if $x > 0$. Hence $\ln y > 0$ if and only if $y > 1$. Therefore $y_1 > y_2 > 0$ implies that

$$\ln y_1 = \ln\left(y_2 \cdot \frac{y_1}{y_2}\right) = \ln y_2 + \ln\left(\frac{y_1}{y_2}\right) > \ln y_2.$$

Finally, differentiate the equation

$$y = e^x = e^{\ln y}$$

with respect to y, one has, by chain rule

$$1 = e^x \cdot \frac{dx}{dy} = y \cdot \frac{dx}{dy}$$

namely

$$\frac{d}{dy} \ln y = \frac{1}{y}.$$

$\square$

Remark. The polynomial functions are solutions of differential equations of the type $D^n y = 0$, the exponential function $y = a^x = e^{kx}$, $k = \ln a$, is a solution of $(D - k)y = 0$ and the trigonometric function $\sin x$ and $\cos x$ are solutions of $(D^2 + 1) \cdot y = 0$. Conversely, the above brief discussion proves that solutions of differential equations of the type $p(D) \cdot y = 0$ can always be expressed as simple combinations of the above three types of elementary functions.

Exercises

1. Let $f(x)$ be an arbitrary polynomial. Show that $(f(x) - f(a))$ is divisible by $(x - a)$. In particular, $f(x)$ is itself divisible by $(x - a)$ if and only if $f(a) = 0$.

2. If $a_1, a_2, \ldots, a_k$ are distinct roots of a given polynomial $f(x)$, then $f(x)$ is divisible by

$$(x - a_1)(x - a_2) \ldots (x - a_k).$$

[Hint: Using Exercise 1 and induction on k.]

3. Show that a polynomial of degree n has at most n distinct roots.

4. Show that a polynomial function $f(x)$ of degree at most n is uniquely determined by its values at $(n + 1)$ points.

5. Find the unique polynomial of degree k whose values at $0, 1, 2, \ldots, k$ are equal to $0, \ldots, 0, 1$.

6. Find the quadratic polynomials $f_1(x)$, $f_2(x)$, $f_3(x)$ and $f(x)$ whose values at a_1, a_2, a_3 are given by the following chart, namely

x	a_1	a_2	a_3
f_1	b_1	0	0
f_2	0	b_2	0
f_3	0	0	b_3
f	b_1	b_2	b_3

7. Recall that $\tan x = \frac{\sin x}{\cos x}$, $\frac{d}{dx}\tan x = ?$
8. Replace the variable x in the power series of e^x, namely

$$e^x = 1 + x + \frac{x^2}{2!} + \cdots + \frac{x^k}{k!} + \cdots$$

by ix and compare the resulting real (resp. imaginary) part with the power series of $\cos x$ (resp. $\sin x$). (This was most likely the way that Euler discovered the formula $e^{ix} = \cos x + i\sin x$.)
9. Show that every solution of $(D^2 + 1)y = 0$ can be expressed uniquely as $c_1 \cos x + c_2 \sin x$ for suitable constants c_1 and c_2.
10. Show that every solution of $(D^2 + 1)^n y = 0$ can be expressed uniquely as

$$f_1(x)\cos x + g_1(x)\sin x$$

for suitable polynomials $f_1(x)$ and $g_1(x)$ of degree at most $(n-1)$.

---

§ 2. Typical Examples of Applications of Calculus

Calculus is a powerful basic tool for mathematical analysis. However, the idea of *mathematical analysis* had already played an important role in human civilization since antiquity. Therefore, for the purpose of providing a historical perspective and illustrating the basic idea of mathematical analysis, let us begin our discussion by an outstanding historical example which, in fact, only uses elementary geometry.

Example 1 (An estimation of the size of the earth by Eratosthenes). Although the local geography of the places that we live may have a variety of unevenness such as mountains, canyons, valleys, etc., the global shape of the earth that we share with many other creatures is almost a perfect round sphere. Therefore, to have an estimate of correct order of magnitude on its size is, of course, a problem of fundamental importance toward the understanding of the world. It was an amazing achievement of human civilization that such an estimate had already been obtained in the 3rd century B.C. by Eratosthenes of Cyrene (284–192 B.C., director of the library of Alexandria). His estimate was based upon the following two pieces of geographic knowledge, namely

(i) At noon of the summer solstice, the sunlight was observed to shine directly down to the bottom of a deep well at Syene (modern Aswan).

(ii) While in Alexandria, which is within 1° of meridian due north of Syene, the sunlight made an angle roughly equal to 1/50 of 360°.

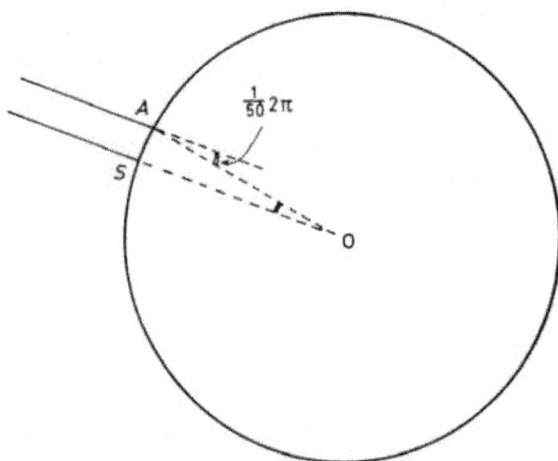

Fig. 36

In applying mathematics to analyze the above geographic facts, he drew a simple diagram (cf. Fig. 36) to represent the "common" meridian circle passing through Syene and Alexandria by a circle and the sunlights at noon of the summer solstice at the above two places by two parallel lines (the reason being that the sun is known to be far, far away as compared to the size of the earth).

Since the sunlight at S (Syene) is perpendicular to the surface, it is pointing directly toward the center O of the circle, while the sunlight at A (Alexandria) makes an angle roughly equal to 1/50 of 360° with the direction of $\overline{AO}$. Thus he concluded that the central angle $\angle AOS$ must be also roughly equal to 1/50 of 360° and hence the circumference of the meridian circle (i.e., the earth) is 50 times the distance between Alexandria and Syene. He estimated the latter simply by the fact that camel caravans, which usually travelled 100 stadia a day, took 50 days to reach Syene from Alexandria. Hence the circumference of the earth is roughly equal to

$$50 \times 50 \times 100 \text{ stadia} \tag{77}$$

which is quite close to the modern measurement of 40 000 km.

Remark. This is a masterpiece of mathematical analysis and a brilliant example of mathematical abstraction. The simple picture of Fig. 36 concisely organized the geographic events at Alexandria and Syene into a well-understood geometric configuration, thus enabled him to deduce the crucial correlation, namely, that the circumference of the earth is roughly 50 times the distance between Alexandria and Syene.

2.1. *Area, volume and gravitation force*

To provide some historical background on the development of integration theory, we discuss here a few pertinent examples which are, in fact, the prelude to modern integral calculus.

Example 2 (Archimedes proof of the area formula of the sphere). Inspired by the monumental success of Eudoxus in applying the approximation methodology to rebuild the foundation of quantitative geometry, geometers of the antiquity were energetically extending the method of approximation to study the areas and volumes of curved objects. They called it the *method of exhaustion*. Archimedes learned this method of exhaustion during his visit to Alexandria and soon became the leading master of this method. Archimedes (287–212 B.C.), reputedly to be the greatest scientist and mathematician of the antiquity, had many outstanding achievements both in science and in geometry. Among his achievements, the one that he certainly prided the most was his proof of the area formula of the sphere, namely

Archimedes Formula. The surface area of a sphere of radius R is equal to the surface area of a circular cylinder of radius R and height $2R$ which is equal to four times the area of a circle of radius R, namely

$$4\pi R^2.$$

This is the famous formula whose geometric proof (cf. Fig. 37) was engraved on his tombstone in accordance with his own wishes.

As indicated in Fig. 37, Archimedes placed the sphere inside the circular cylinder and then cut both into very, very thin strips by horizontal parallel planes.

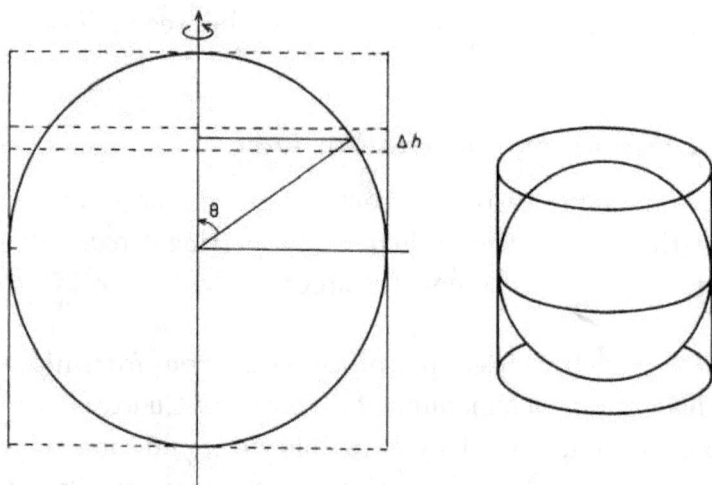

Fig. 37

Then he applied the principle of exhaustion (i.e., approximation) to show that the corresponding thin strips are always of the same area, thus proving that they have the same total area. In terms of modern terminology, the cylindrical strip is of length $2\pi R$ and of width Δh, while the spherical strip is approximately of length $2\pi R \sin\theta$ and of width $\Delta h/\sin\theta$.

Example 3 (The volume formula of a cone). The volume of a cone with base area A and height h is equal to

$$\frac{1}{3}h \cdot A.$$

This formula was again proved by the method of exhaustion in the antiquity. The following is its modern adaptation.

Proof. Choose the vertex of the cone as the origin O and the direction perpendicular to the base of the cone as the z-axis. Then the cross-section area of the plane $z = z_0$, $0 \le z_0 \le h$ is equal to

$$A \cdot \left(\frac{z_0}{h}\right)^2.$$

(The cross-section is a similar planar region of the base with linear magnification factor of $\frac{z_0}{h}$.) Therefore

$$V = \int_0^h A \cdot \left(\frac{z}{h}\right)^2 dz = \frac{A}{h^2} \int_0^h z^2 dz$$
$$= \frac{A}{h^2} \cdot \frac{h^3}{3} = \frac{1}{3}hA.$$

$\square$

Example 4 (Newton's formula of the gravitation force of a spherical symmetric body). Although celestial bodies such as the sun and the planets are huge balls, the distances between them are immensely larger than their sizes. Therefore, it is physically still reasonable to treat them as mere particles in analyzing the forces causing the planets to move in accordance to the three Kepler laws. However, Newton realized that he definitely needed a formula to compute the gravitation force between the earth and an object such as a human body or an apple, in order to demonstrate that it is, indeed, the same kind of force as the one implied by Kepler's law of planet motions. The simplest and best possible formula would be that the composite gravitation force acting on an outside particle by a body with *spherically symmetric* mass- distribution is equal to the force if all the mass of a ball is concentrated at the center. If this can be proved to be the case, then such a formula will be a nice simple formula for analyzing the total gravitation force that we and everything else on the earth are subject to all of the time. This is, of course, a natural problem for Newton to apply the technique of integration. However, Newton was trying hard to provide a purely geometric proof of such a formula in order to conform with the overall spirit and his book Principia. According to Arnold, the difficulty of producing such a proof was one of the reasons that delayed its publication. Indeed, Newton's geometric proof of this formula in Principia was a rather difficult one. However, one naturally expects that there should be a nice, simple geometric proof of such a nice formula for such a perfect shape! We include here a simple geometric proof of Newton's formula to substantiate the belief that perfect spherical symmetry, somehow, always harbors neat proof!

A Simple Proof of Newton's Formula. The proof can easily be reduced to the case of a thin spherical shell with uniform density per unit area. The geometric setting is a given sphere and an outside point P. For such a geometry, the *spherical conjugate point*, P', of P which is, by definition, the point on $\overline{OP}$ with $\overline{OP'} \cdot \overline{OP} = R^2$ is a rather *special* point.

The key idea of the simple proof is to use the subdivision of the spherical surface induced from the subdivision of the total solid angle *centered* at this special point P'. As indicated in Fig. 38,

$$\Delta OPQ \sim \Delta OQP', \quad dA = \overline{P'Q}^2 \frac{d\sigma}{\cos\theta}$$

and the contribution of $d\mathbb{F}$ toward the total composite force is its component toward the center, namely, $|d\mathbb{F}|\cos\theta$. Therefore

$$|d\mathbb{F}| \cdot \cos\theta = G \frac{m_1 \cdot \rho dA}{\overline{PQ}^2}\cos\theta$$

$$= Gm_1\rho \frac{\overline{P'Q}^2}{\overline{PQ}^2} d\sigma = Gm_1\rho \frac{R^2}{\overline{OP}^2} d\sigma \; .$$

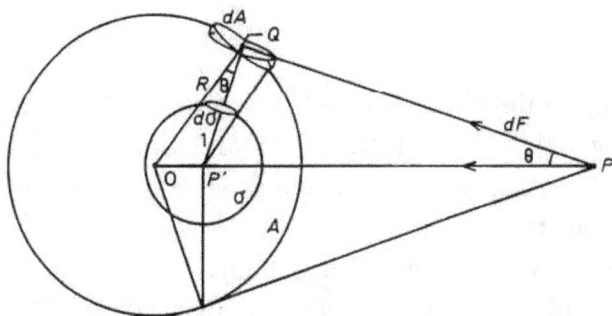

Fig. 38

Hence, the total force is equal to

$$\int |d\mathbb{F}| \cdot \cos\theta = \int Gm_1\rho \frac{R^2}{\overline{OP}^2} d\sigma$$

$$= Gm_1\rho \frac{R^2}{\overline{OP}^2} \int d\sigma = G\frac{m_1 4\pi R^2 \rho}{\overline{OP}^2} = G\frac{m_1 m_2}{\overline{OP}^2} \; .$$

$\square$

2.2. *Derivatives and extremals*

The derivative $f'(x)$ of a given function $f(x)$ records the *rate of change* of $f(x)$ wherever it is defined. For most functions occurring in applications, their derivatives are continuous functions except possibly for some isolated points. If $f(x)$ has a local maximal or minimal at $x = a$ and $f'(a)$ is defined, then it is quite obvious that $f'(a)$ must be zero! Therefore, the studying of the zeros of $f'(x)$ and those possible isolated points that $f'(x)$ *cannot* be defined provides an effective way of determining the extremals of a given function $f(x)$. Anyhow, differentiation is a powerful tool for analyzing the extremals of a given function and *optimization* plays an important role in natural phenomena.

Example 5 (Reflection of light). The following well-known law of reflection is, of course, the result of many experimental measurements, namely

Law of Reflection. To each point on the surface of reflection, the entering light, the reflected light and the normal to the surface are always coplanar, and moreover, the normal bisects the angle between the former two.

Now, we shall apply mathematical analysis to seek a deeper understanding of the geometric or physical meaning of the above law of reflection.

Analysis

(i) Since the entering light, the reflected light and the normal are coplanar, it is convenient to restrict our discussion to that plane, thus simplifying the geometric setting without loss of generality.

(ii) Let us first consider the simplest case that the surface of reflection is actually a plane. In this case, the restricted planar setting is indicated in Fig. 39.

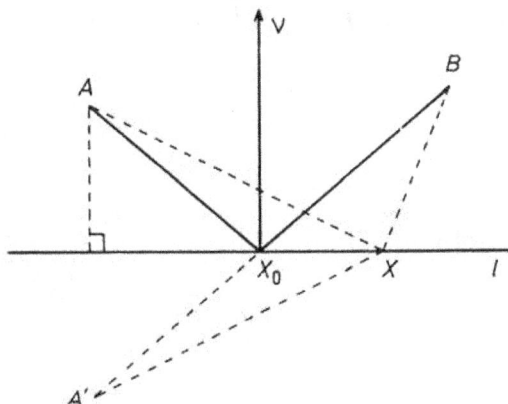

Fig. 39

Let A' be the symmetric point of A. Then

$$\overline{AX} = \overline{A'X} \tag{78}$$

for any point X in l (or in the plane). Hence

$$\overline{AX} + \overline{XB} = \overline{A'X} + \overline{XB} \geq \overline{A'B} = \overline{A'X_0} + \overline{X_0B} = \overline{AX_0} + \overline{X_0B} \tag{79}$$

and equality holds only when $X = X_0$. Therefore, the geometric meaning here is that light travels the *shortest pathway* linking a pair of given points A and B *via* a point on the plane of reflection.

(iii) Finally, let us consider the general case that the surface of reflection is a piece of smooth surface. Choose a given point X_0 on the surface as the origin, the plane containing the three lines as the (x,y)-plane and the normal as the y-axis. Then the intersection of the surface of reflection and the (x,y)-plane can be locally represented by

$$y = f(x), \ f(0) = 0 \text{ and } f'(0) = 0. \tag{80}$$

Let $A(a_1, a_2)$ and $B(b_1, b_2)$ be fixed points on the entering line and the reflected line at the origin respectively and $P(x, f(x))$ be a nearby point on the curve. Then

$$\begin{aligned}
l(x) &= \overline{AP} + \overline{PB} \\
&= \sqrt{(x - a_1)^2 + (f(x) - a_2)^2} + \sqrt{(x - b_1)^2 + (f(x) - b_2)^2}.
\end{aligned} \tag{81}$$

Differentiate (81) and use $f(0) = f'(0) = 0$, one gets

$$l'(0) = \frac{-a_1}{\sqrt{a_1^2 + a_2^2}} + \frac{-b_1}{\sqrt{b_1^2 + b_2^2}} = \sin\alpha + \sin\beta = 0. \qquad (82)$$

This shows that $\overline{AX_0} + \overline{X_0B}$ is again a local minimal among $\overline{AP} + \overline{PB}$.

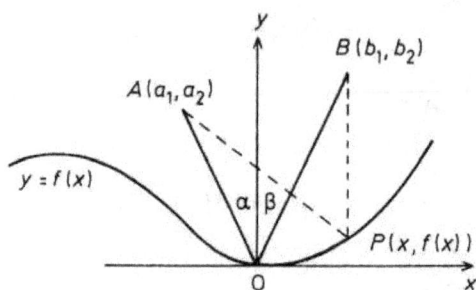

Fig. 40

Example 6 (Snell's law of refraction). When light goes through a smooth surface separating two different kinds of mediums, the Snell's law of refraction asserts that
  (i) The entering light, the refracted light and the normal to the surface are coplanar.
 (ii) The two angles λ (resp. μ) between the normal and the entering light (resp. the refracted light) satisfy the following relationship, namely

$$\sin\mu = c \cdot \sin\lambda \qquad (83)$$

where c is a constant depending on the two types of medium and the wave-length of the light, called the *refraction coefficient.*
(iii) For light of a fixed wavelength, the above refraction coefficient has the following multiplicative property, namely, the refraction coefficient between mediums of the first and the third kind is equal to the *product* of the refraction coefficient between mediums

of the first and the second kind and the refraction coefficient between mediums of the second and third kinds.

Mathematical Analysis of the Snell's Law

1. In the case of refraction, the pathway of light is, in general, not a straight interval and hence it is *no longer* the shortest pathway! What is the physical difference that makes refraction and reflection behave differently? The only difference between them is that light travels in the same kind of medium in the case of reflection, while, in the case of refraction, light travels in two different kinds of medium. Maybe the refraction phenomena is due to the difference of the speeds of light in different mediums. If light travels with uniform speed such as in the case of reflection, then the fastest pathway is, of course, just the shortest pathway. If light travels with different speeds in different regions such as in the case of refraction, then the fastest pathway is, in general, *not* the shortest pathway. Therefore, the physically attractive *hypothesis* that light travels in the fastest pathway may still provide a unified explanation of both reflection and refraction.

2. Anyway, let us again apply calculus to analyze the above "*fastest pathway hypothesis*". Let v_1 (resp. v_2) be the speed of light in medium of the first (resp. second) kind. Then the total time of the pathway $\overline{AP} + \overline{PB}$ is given by

$$T(x) = \frac{\sqrt{(x - a_1)^2 + (f(x) - a_2)^2}}{v_1} + \frac{\sqrt{(x - b_1)^2 + (f(x) - b_2)^2}}{v_2}$$

(84)

Compute $T'(x)$ and make use of the condition $f(0) = f'(0) = 0$. One gets

$$T'(0) = \frac{-a_1}{v_1\sqrt{a_1^2 + a_2^2}} + \frac{-b_1}{v_2\sqrt{b_1^2 + b_2^2}} = \frac{\sin \lambda}{v_1} - \frac{\sin \mu}{v_2} = 0 , \quad (85)$$

which shows that

$$\sin \lambda = \frac{v_1}{v_2} \sin \mu.$$

(86)

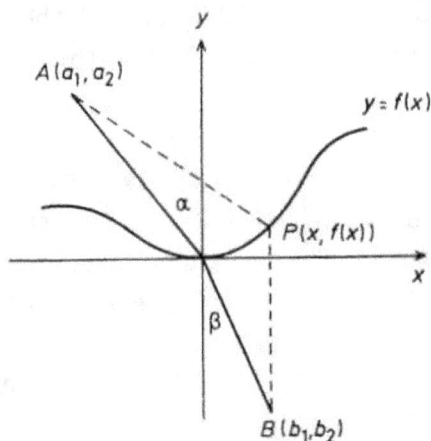

Fig. 41

The above rather simple mathematical analysis not only demonstrates the compatibility between the experimental Snell's law of refraction and the fastest pathway hypothesis but also reveals the neat physical meaning of the refraction coefficient, namely

$$c = \frac{v_1}{v_2}. \tag{87}$$

Notice that the above physical meaning of the refraction coefficient also provides a straightforward explanation of the multiplicative property among refraction coefficients, namely

$$\frac{v_1}{v_2} \cdot \frac{v_2}{v_3} = \frac{v_1}{v_3}. \tag{88}$$

Remark. The above mathematical analysis of experimental laws on reflection and refraction naturally leads to the Fermat's least time principle in geometric optics, namely, light always travels in the fastest pathway. Such a neat principle not only constitutes a far-reaching and deeper understanding of the nature of light, but it also inspired the later formulation of the principle of least action by Maupertius, Euler and Lagrange, and the Hamilton principle.

2.3. *Local approximation and osculating curves*

Example 7 (Osculating circle and curvature). A curve can usually be considered as the locus of a moving point. Thus, it is natural to represent a planar (resp. space) curve by a pair (resp. triple) of functions recording the changing of x and y (resp. x, y and z) coordinates respectively, namely

$$x = f(t), \ y = g(t) \text{ (resp. and } z = \varphi(t)). \tag{89}$$

Let $P_i = (f(t_i), g(t_i))$, $i = 0, 1, 2$, be three points on such a given curve. It is a well-known fact in elementary geometry that three points determine a circle (may be of infinite radius). Using determinant, the equation of the unique circle passing through P_0, P_1 and P_2, say $\Gamma(P_0, P_1, P_3)$, can be written as follows

$$\begin{vmatrix} 1 & X & Y & X^2 + Y^2 \\ 1 & f(t_0) & g(t_0) & h(t_0) \\ 1 & f(t_1) & g(t_1) & h(t_1) \\ 1 & f(t_2) & g(t_2) & h(t_2) \end{vmatrix} = 0 \tag{90}$$

where $h(t) = f(t)^2 + g(t)^2$.

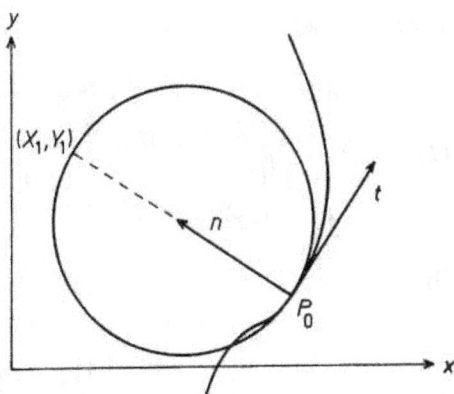

Fig. 42

The osculating circle of the given curve at P_0 is, by definition, the limiting circle of $\Gamma(P_0, P_1, P_2)$ as P_1 and $P_2 \to P_0$. Such a limiting circle exists if both $f(t)$ and $g(t)$ are of C^2-type and the following is a typical example of using the mean value theorem to compute the equation of such a limiting object.

Let us first consider the simpler problem of finding the limiting circle of fixing P_2 but letting $P_1 \to P_0$. Notice that the left-hand side of (90) will become identically zero if one simply replaces t_1 by t_0 because the second and third rows become the same! This means that all the coefficients of (90) tend to zero as their limits as $t_1 \to t_0$. Therefore, one needs to find out the reason and to get rid of the troublemaker before taking the limit. This is exactly where the mean value theorem comes in as the "troubleshooter".

Using the property of determinants, one first rewrites the left-hand side of (90) by subtracting the second row from the third row, namely

$$\begin{vmatrix} 1 & X & Y & X^2 + Y^2 \\ 1 & f(t_0) & g(t_0) & h(t_0) \\ 0 & f(t_0) - f(t_0) & g(t_1) - g(t_0) & h(t_1) - h(t_0) \\ 1 & f(t_2) & g(t_2) & h(t_2) \end{vmatrix} = 0. \qquad (90')$$

Then, by mean value theorem, the left-hand side of (90') can be rewritten as follows, namely

$$(t_1 - t_0) \begin{vmatrix} 1 & X & Y & X^2 + Y^2 \\ 1 & f(t_0) & g(t_0) & h(t_0) \\ 0 & f'(\xi_1) & g'(\eta_1) & h'(\zeta_1) \\ 1 & f(t_2) & g(t_2) & h(t_2) \end{vmatrix} = 0 \qquad (90'')$$

where ξ_1, η_1, ζ_1 lie between t_0 and t_1. Now, it is clear that the factor $(t_1 - t_0)$ is exactly the troublemaker. Get rid of it and then take the limit $t_1 \to t_0$. One gets the equation of the limiting circle which is tangent to the curve at P_0 and passing through P_2, say denoted by $\Gamma(P_0, P_0, P_2)$, namely

$$\begin{vmatrix} 1 & X & Y & X^2 + Y^2 \\ 1 & f(t_0) & g(t_0) & h(t_0) \\ 0 & f'(t_0) & g'(t_0) & h'(t_0) \\ 1 & f(t_2) & g(t_2) & h(t_2) \end{vmatrix} = 0. \tag{91}$$

Next let us consider the problem of $\lim_{P_2 \to P_0} \Gamma(P_0, P_0, P_2)$. Clearly, we again need to do some fixing of (91) before taking the limit of $t_2 \to t_0$ lest all of the coefficients of (91) will tend to zero. In fact, exactly the same kind of fixing is no longer sufficient. What is actually needed is to replace the fourth row of (91) by

$$0, f(t_2) - f(t_0) - (t_2 - t_0)f'(t_0),$$
$$g(t_2) - g(t_0) - (t_2 - t_0)g'(t_0), h(t_2) - h(t_0) - (t_2 - t_0)h'(t_0) \tag{92}$$

namely, by Taylor's theorem

$$0, f''(\xi_2)\frac{(t_2 - t_0)^2}{2}, g''(\eta_2)\frac{(t_2 - t_0)^2}{2}, h''(\zeta_2)\frac{(t_2 - t_0)^2}{2}. \tag{92'}$$

Thus obtaining the following modified form of (91), namely

$$\frac{(t_2 - t_0)^2}{2} \begin{vmatrix} 1 & X & Y & X^2 + Y^2 \\ 1 & f(t_0) & g(t_0) & h(t_0) \\ 0 & f'(t_0) & g'(t_0) & h'(t_0) \\ 0 & f''(\xi_2) & g''(\eta_2) & h''(\zeta_2) \end{vmatrix} = 0 \tag{91'}$$

where ξ_2, η_2, ζ_2 lie between t_0 and t_2. Again get rid of the trouble-maker factor of $\frac{(t_2 - t_0)^2}{2}$ and then take the limit of $t_2 \to t_0$. One obtains the equation of the limiting circle, namely

$$\begin{vmatrix} 1 & X & Y & X^2 + Y^2 \\ 1 & f(t_0) & g(t_0) & h(t_0) \\ 0 & f'(t_0) & g'(t_0) & h'(t_0) \\ 0 & f''(t_0) & g''(t_0) & h''(t_0) \end{vmatrix} = 0. \tag{93}$$

This proves the existence of the osculating circle of a C^2-curve at a given point P_0 together with the explicit determination of its equation.

Geometrically speaking, circles are curves with uniform curvatures which are equal to $1/R$. Thus, the curvature of a C^2-curve at a given point is defined to be the curvature of its osculating circle at the given point. Now let us proceed to compute the radius of the above osculating circle given by (93).

Since the radius of a circle is clearly invariant under translation, we may assume without loss of generality that $f(t_0) = g(t_0) = 0$. Then (93) simplifies into

$$\begin{vmatrix} 1 & X & Y & X^2 + Y^2 \\ 1 & 0 & 0 & 0 \\ 0 & f'(t_0) & g'(t_0) & 0 \\ 0 & f''(t_0) & g''(t_0) & 2\Delta^2 \end{vmatrix} = 0 \tag{93'}$$

where $\Delta = \sqrt{f'(t_0)^2 + g'(t_0)^2}$.

Let $\mathbf{t}$ and $\mathbf{n}$ be the unit tangent and normal vector at P_0, it is easy to see that

$$\mathbf{t} = \left(\frac{f'(t_0)}{\Delta}, \frac{g'(t_0)}{\Delta} \right), \quad \mathbf{n} = \left(-\frac{g'(t_0)}{\Delta}, \frac{f'(t_0)}{\Delta} \right). \tag{94}$$

As indicated in Fig. 42,

$$(X_1, Y_1) = \left(-\frac{2Rg'(t_0)}{\Delta}, \frac{2Rf'(t_0)}{\Delta} \right) \tag{95}$$

is also a point on the osculating circle. Hence

$$\begin{vmatrix} 1 & -\frac{2R}{\Delta}g'(t_0) & \frac{2R}{\Delta}f'(t_0) & 4R^2 \\ 1 & 0 & 0 & 0 \\ 0 & f'(t_0) & g'(t_0) & 0 \\ 0 & f''(t_0) & g''(t_0) & 2\Delta^2 \end{vmatrix} = 0. \tag{96}$$

From (96), it is straightforward to show that

$$\frac{1}{R} = \frac{f'(t_0)g''(t_0) - g'(t_0)f''(t_0)}{\Delta^3}. \tag{97}$$

This is the curvature formula of a plane curve.

Example 8 (Higher order osculating polynomial curves of the graph $y = f(x)$). As it has already been discussed in Sec. 1, the concept of derivative and higher order derivatives is closely related to the local approximation of a given function $f(x)$ by polynomial functions. Roughly speaking, a C^k-function can be locally approximated by a polynomial function (of degree at most k) up to order k. Geometrically, this corresponds to the existence of an osculating polynomial curve of higher order of contact.

Let $\{P_i(x_i, f(x_i)); \ 0 \leq i \leq k\}$ be $(k+1)$ points on the graph $y = f(x)$ of a C^k-function. Then there exists a unique polynomial of degree at most k whose graph passes through the above $(k+1)$ points. Fix x_0 and let the other x_i, $1 \leq i \leq k$, all tend to x_0 as limit. Then the limiting curve is called the osculating polynomial curve of order k at P_0. In fact, this osculating polynomial curve is exactly

$$y = f(x_0) + f'(x_0)(x - x_0) + \frac{f''(x_0)}{2!}(x - x_0)^2 + \cdots + \frac{f^{(k)}(x_0)}{k!}(x - x_0)^k. \tag{98}$$

Proof. For simplicity of notation and computation, one may assume that $x_0 = 0$ without loss of generality. Again, using determinant, the polynomial graph of degree at most k can be represented by

$$\begin{vmatrix} 1 & x & x^2 & \cdots & x^k & y \\ 1 & 0 & 0 & & 0 & f(0) \\ 1 & x_1 & x_1^2 & & x_1^k & f(x_1) \\ \cdots & \cdots & \cdots & \cdots & \cdots & \cdots \\ 1 & x_k & x_k^2 & & x_k^k & f(x_k) \end{vmatrix} = 0. \tag{99}$$

The same kind of computation using Taylor's theorem as that of the osculating circle enables us to prove that the osculating polynomial curve of order k can be represented by the following equation, namely

$$
\begin{vmatrix}
1 & x & x^2 & \cdots & x^k & y \\
1 & 0 & 0 & & 0 & f(0) \\
0 & 1 & 0 & & 0 & f'(0) \\
0 & 0 & 2 & \ddots & \vdots & f''(0) \\
\vdots & \vdots & \vdots & \vdots & \vdots & \vdots \\
& & & 0 & & \vdots \\
0 & 0 & \cdots & 0 & k! & f^{(k)}(0)
\end{vmatrix} = 0.
\tag{100}
$$

The expansion of the above equation is as follows,

$$
(1 \cdot 2! \cdot 3! \ldots k!) \cdot \left[y - f(0) - f'(0)x - f''(0)\frac{x^2}{2!} - \cdots - f^{(k)}(0)\frac{x^k}{k!} \right] = 0.
\tag{100$'$}
$$

Hence, the osculating polynomial curve of order k is the graph of

$$
y = f(0) + f'(0)x + \frac{f''(0)}{2!}x^2 + \cdots + \frac{f^{(k)}(0)}{k!}x^k.
\tag{100$''$}
$$

$\square$

2.4. *Laws of nature and differential equations*

Example 9 (Population growth and radioactive decay). If the growth rate of a population is kept at a certain constant k, then the total population as a function of time satisfies the differential equation

$$
Dy = ky \quad \text{or} \quad (D - k)y = 0.
\tag{101}
$$

Hence

$$
y = c \cdot e^{kt}
\tag{102}
$$

where c is the initial population. In particular,

$$T = \frac{\ln 2}{k} \tag{103}$$

is called the doubling time.

Similarly the decay rate of a given radioactive substance is essentially a negative growth rate. Hence the amount of a given radioactive material as a function of time satisfies the differential equation

$$Dy = ky \text{ or } (D - k)y = 0 \tag{101'}$$

with negative k. Hence

$$y = c \cdot e^{kt}, \; k < 0 \tag{102'}$$

where c is the initial amount of the radioactive material. In particular,

$$L = \frac{\ln 2}{-k} \tag{103'}$$

is called the half-life of that radioactive material.

Example 10 (Hooke's law of elasticity and simple harmonic motion). Let us first consider a simple mechanical system consisting of a weight of mass m suspended by a pair of balanced spring, cf. Fig. 43.

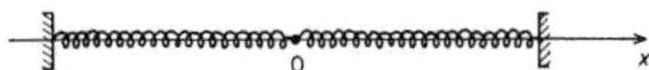

Fig. 43

Choose the equilibrium position to be the origin of the x-axis. Then, by Hooke's law, the restoration force at the position x is equal to $-kx$. Hence

$$m \cdot \frac{d^2x}{dt^2} = -kx \text{ or } \left(D^2 + \frac{k}{m}\right) \cdot x = 0 \tag{104}$$

where k is the elastic constant of the system. Therefore, by Theorem 5.5' and Exercise 9 of Sec. 1.3,

$$x = c_1 \cos \sqrt{\frac{k}{m}} t + c_2 \sin \sqrt{\frac{k}{m}} t \tag{105}$$

where c_1 and c_2 are suitable constants depending on the initial position and initial velocity as follows, namely

$$x(0) = c_1 \qquad \left. \frac{dx}{dt} \right|_{t=0} = c_2 \sqrt{\frac{k}{m}}.$$

Motions described by (105) are called simple harmonic motions. They are always periodic and the period is equal to

$$T = 2\pi \sqrt{\frac{m}{k}}. \tag{106}$$

Set

$$\rho = \sqrt{c_1^2 + c_2^2} \qquad \varphi = \tan^{-1} \frac{c_2}{c_1}. \tag{107}$$

Then

$$x = \rho \cdot \left[\cos\varphi \cos \sqrt{\frac{k}{m}} t + \sin\varphi \sin \sqrt{\frac{k}{m}} t \right]$$

$$= \rho \cos\left(\sqrt{\frac{k}{m}} t - \varphi \right). \tag{108}$$

Remark. ρ is called the amplitude and φ is called the phase angle of the simple harmonic motion of (105). The geometric meaning of (108) is that the simple harmonic motion is, in fact, the x-component of the circular motion on the circle

$$x^2 + y^2 = \rho^2$$

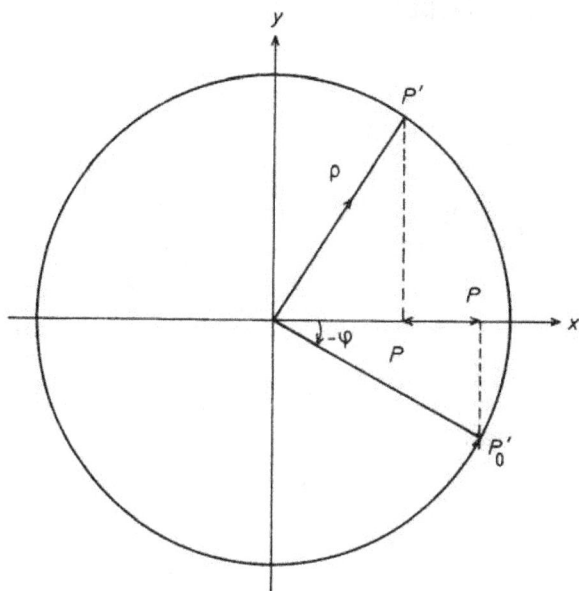

Fig. 44

with constant angular velocity equal to $\sqrt{\frac{k}{m}}$ and initial angle equal to $-\varphi$, cf. Fig. 44.

Example 11 (Mathematical analysis of the three Kepler's laws of planet motions). At the beginning of the seventeenth century, J. Kepler discovered three laws of planet motions based upon the remarkably accurate data of astronomical measurement accumulated by the lifetime devotion of Tycho de Brahe. This revolutionary discovery of Kepler was the monumental breakthrough which marked the beginning of modern science.

Kepler's laws of planet motions

First law: Each planet moves around the sun in an elliptical orbit with the sun at one of its foci.
Second law: The area swept out per unit time by the segment joining the sun toward the planet, as indicated in Fig. 45, is always a constant.

Third law: The ratios between the cube of the major axes of the elliptical orbits and the square of the periods for *all planets in the solar system* are equal to each other.

The mathematical analysis together with all the twists and turns that eventually led him to the discovery of the above three laws was fully recorded in his own books which are, forever, masterpieces in both scientific research and mathematical analysis. However, what we are going to discuss here is the mathematical analysis on the physical content of Kepler's laws.

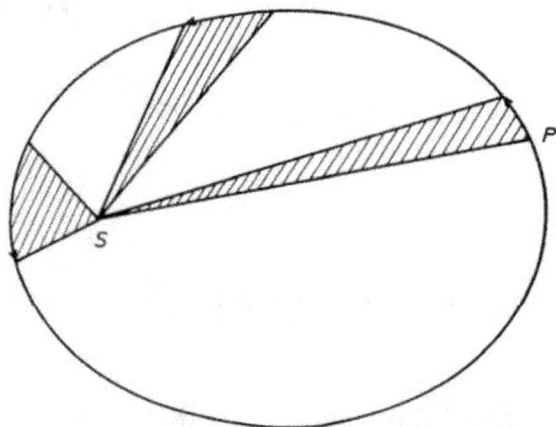

Fig. 45

Analysis

(i) Let us first analyze the physical meaning of Kepler's second law. Naturally, it is convenient to use the polar coordinate system with the position of the sun at the origin. Let the position of a given planet be $P(r, \theta)$. Then, when the time changes from t to $t + dt$, the area swept out by the segment $\overline{OP}$ is given by

$$dA = \frac{1}{2}r^2 d\theta. \tag{109}$$

Hence, the second law states that

$$\frac{dA}{dt} = \frac{1}{2}r^2\frac{d\theta}{dt} = \frac{1}{2}r^2 w = \text{ constant} \tag{110}$$

where $w = \frac{d\theta}{dt}$ is usually called the *angular velocity*. Let m be the mass of the planet. Then $m \cdot r^2 w$ is called the *angular momentum*, and hence, Kepler's second law simply asserts the *conservation of angular momentum* in the case of planet motion. The following are the general definitions of angular momentum and torque in physics, namely

Definitions. Let $\mathbb{F}$ be a force acting on a particle P and O be the fixed center. Then the *torque* of $\mathbb{F}$ with respect to O is defined to be the vector $\overrightarrow{OP} \times \mathbb{F}$. Let $\mathbf{v}$ be the velocity vector of the moving particle P of mass m. Then $m\mathbf{v}$ is the momentum and $\overrightarrow{OP} \times m\mathbf{v}$ is called the *angular momentum* with respect to O. (The $\times$-product of two vectors $\mathbf{a}$ and $\mathbf{b}$, denoted by $\mathbf{a} \times \mathbf{b}$, is characterized by the property that

$$(\mathbf{a} \times \mathbf{b}) \cdot \mathbf{c} \equiv \det(\mathbf{a}, \mathbf{b}, \mathbf{c}). \tag{111}$$

If $\mathbf{a} = (a_1, a_2, a_3)$ and $\mathbf{b} = (b_1, b_2, b_3)$, then

$$\mathbf{a} \times \mathbf{b} = \left(\begin{vmatrix} a_2 & a_3 \\ b_2 & b_3 \end{vmatrix}, \begin{vmatrix} a_3 & a_1 \\ b_3 & b_1 \end{vmatrix}, \begin{vmatrix} a_1 & a_2 \\ b_1 & b_2 \end{vmatrix} \right). \tag{112}$$

Moreover,

$$(\mathbf{a} \times \mathbf{b}) \cdot (\mathbf{c} \times \mathbf{d}) = \begin{vmatrix} \mathbf{a} \cdot \mathbf{c}, & \mathbf{b} \cdot \mathbf{c} \\ \mathbf{a} \cdot \mathbf{d}, & \mathbf{b} \cdot \mathbf{d} \end{vmatrix} \tag{113}$$

is also an important identity in vector algebra which can be regarded as the Pythagoras Theorem of areas.)

Basic Fact

$$\frac{d}{dt}(\overrightarrow{OP} \times m\mathbf{v}) = \mathbf{v} \times m\mathbf{v} + \overrightarrow{OP} \times m\frac{d}{dt}\mathbf{v}$$

$$= \overrightarrow{OP} \times \mathbb{F} \tag{114}$$

Therefore, the conservation of angular momentum is equivalent to the condition

$$\overrightarrow{OP} \times \mathbf{F} \equiv 0 \tag{115}$$

namely, $\overrightarrow{OP}$ and $\mathbf{F}$ are *colinear*. Hence, the physical meaning of the second law is that $\overrightarrow{OP}$ and $\mathbf{F}$ are always collinear, namely, $\mathbf{F}$ is in the radial direction.

(ii) Next let us analyze the physical meaning of the first law. Let us begin with a brief review on the geometry of ellipse. It is well-known that an ellipse with a and b as its major and minor axes can be represented by the following simple form of equation, namely

$$\frac{x^2}{a^2} + \frac{y^2}{b^2} = 1. \tag{116}$$

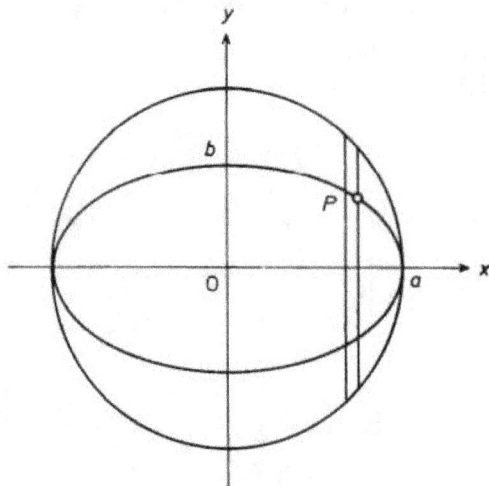

Fig. 46

The coordinates of its two foci are $(\pm c, 0)$ where $c = \sqrt{a^2 - b^2}$. As indicated in Fig. 46, the length of each strip of width dx is always equal to $\frac{b}{a}$ times of the corresponding strip of the circle of radius a. Therefore, the area of the ellipse is equal to $\frac{b}{a}$ times πa^2, which is equal to πab.

Notice that the origin 0 for the polar coordinate system is *not* the same as the origin for the cartesian coordinate system; its cartesian coordinate is $(-c, 0)$. The coordinate transformation between (r, θ) and (x, y) are given as follows, namely

$$x = r \cos \theta - c, \ y = r \sin \theta. \tag{117}$$

Substitute the above expression of x and y into (46). One gets

$$b^2 (r \cos \theta - c)^2 + a^2 r^2 \sin^2 \theta^2 - a^2 b^2 = 0 \tag{118}$$

namely

$$a^2 r^2 - (r^2 c^2 \cos^2 \theta + 2rc \cos \theta b^2 + b^4) = 0 \tag{118'}$$

or

$$ar \pm (rc \cos \theta + b^2) = 0. \tag{118''}$$

It is easy to see that the choice should be

$$ar - rc \cos \theta - b^2 = 0 , \tag{119}$$

or

$$r = \frac{b^2}{a(1 - e \cos \theta)}, e = \frac{c}{a}. \tag{119'}$$

This is the equation of the ellipse in a polar coordinate. Since the second law asserts that $\frac{1}{2} r^2 w = $ constant,

$$\pi ab = \int_0^T dA = \int_0^T \frac{1}{2} r^2 w \, dt = \frac{1}{2} r^2 w T \tag{120}$$

where T is the *period*. Therefore

$$\frac{2\pi ab}{T} = r^2 w, \ \left(\frac{2\pi ab}{rT}\right)^2 = (rw)^2 \tag{121}$$

and hence, by (119') and (121)

$$rw^2 = \frac{4\pi^2 a^3}{T^2} \frac{1}{r^2} (1 - e \cos \theta). \tag{122}$$

Let $\mathbf{a}$ be the *acceleration* vector of the motion of P and $\mathbf{a}_r$ and $\mathbf{a}_\theta$ be the radial and angular component of $\mathbf{a}$. By (i), $\mathbf{a}_\theta = 0$ and hence $\mathbf{a} = \mathbf{a}_r$. We recall here a basic general formula of $\mathbf{a}_r$, namely

Basic Formula for $\mathbf{a}_r$

$$\mathbf{a}_r = \left(\frac{d^2 r}{dt^2} - rw^2 \right) \mathbf{e}_r, \quad \mathbf{e}_r = (\cos\theta, \sin\theta). \qquad (123)$$

Proof.

$$\overrightarrow{OP} = (r\cos\theta, r\sin\theta)$$

$$\mathbf{a} = \frac{d^2}{dt^2}\overrightarrow{OP} = \left(\frac{d^2}{dt^2}(r\cos\theta), \frac{d^2}{dt^2}(r\sin\theta) \right)$$

$$\frac{d^2}{dt^2}(r\cos\theta) = \frac{d}{dt}\left(\frac{dr}{dt}\cos\theta - rw\sin\theta \right)$$

$$= \left(\frac{d^2 r}{dt^2} - rw^2 \right)\cos\theta - \left(\frac{dr}{dt}w + \frac{d}{dt}(rw) \right)\sin\theta \qquad (124)$$

$$\frac{d^2}{dt^2}(r\sin\theta) = \frac{d}{dt}\left(\frac{dr}{dt}\sin\theta + rw\cos\theta \right)$$

$$= \left(\frac{d^2 r}{dt^2} - rw^2 \right)\sin\theta + \left(\frac{dr}{dt}w + \frac{d}{dt}(rw) \right)\cos\theta.$$

Therefore

$$\mathbf{a} = \left(\frac{d^2 r}{dt^2} - rw^2 \right)\cdot\mathbf{e}_r + \left(\frac{dr}{dt}w + \frac{d}{dt}(rw) \right)\mathbf{e}_\theta$$

$$\mathbf{e}_r = (\cos\theta, \sin\theta), \quad \mathbf{e}_\theta = (-\sin\theta, \cos\theta) \qquad (125)$$

Now, let us compute $\frac{d^2 r}{dt^2}$ as follows. First rewrite Eq. (119') as

$$\frac{1}{r} = \frac{a}{b^2}(1 - e\cos\theta).$$

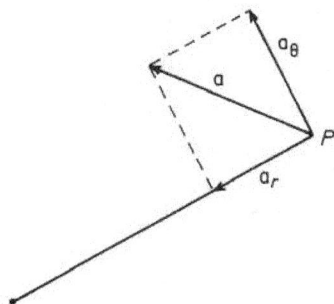

Fig. 47

Therefore, by differentiation with respect to t,

$$-\frac{\frac{dr}{dt}}{r^2} = \frac{a}{b^2} e \sin \theta w. \tag{126}$$

Hence

$$-\frac{dr}{dt} = \frac{a}{b^2} e \sin \theta r^2 w = \frac{2\pi ab}{T} \frac{a}{b^2} e \sin \theta. \tag{127}$$

Differentiate again with respect to t. One gets

$$\begin{aligned}
-\frac{d^2 r}{dt^2} &= \frac{2\pi ab}{T} \frac{a}{b^2} e \cos \theta \cdot w = \frac{2\pi ab}{T} \frac{a}{b^2} e \cos \theta \cdot \frac{r^2 w}{r^2} \\
&= \left(\frac{2\pi ab}{T}\right)^2 \frac{ae \cos \theta}{b^2 r^2} = \frac{4\pi^2 a^3}{T^2} \frac{e \cos \theta}{r^2}.
\end{aligned} \tag{128}$$

Therefore

$$\frac{dr^2}{dt^2} - rw^2 = -\frac{4\pi^2 a^3}{T^2} \frac{1}{r^2} \tag{129}$$

namely

$$\mathbf{a}_r = -\left(\frac{dr^2}{dt^2} - rw^2\right) \mathbf{e}_r = -\frac{4\pi^2 a^3}{T^2} \frac{1}{r^2} \cdot \mathbf{e}_r \tag{130}$$

and

$$\mathbb{F} = m \cdot \mathbf{a} = m\mathbf{a}_r = -\frac{4\pi^2 a^3}{T^2} \cdot m \cdot \frac{1}{r^2} \cdot \mathbf{e}_r. \tag{130'}$$

The above mathematical analysis demonstrates that the "force" that causes the planet to move around the sun is a vector pointing toward the sun with its magnitude *inversely proportionate* to the *square of the distance* between them, namely

$$|\mathbb{F}| = \frac{4\pi^2 a^3}{T^2} m \cdot \frac{1}{r^2}. \tag{130''}$$

Moreover, the third law asserts that the above coefficient $4\pi^2 a^3/T^2$ is the *same* for all the planets of the solar system! Therefore, it is natural to rewrite the above formula in a more symmetric form, namely

$$|\mathbb{F}| = G\frac{Mm}{r^2} \tag{131}$$

where M is the mass of the sun and $G = \frac{4\pi^2 a^3}{T^2 M}$.

The above mathematical analysis constitutes the major step toward the formulation of the *universal gravitation law of Newton* which, of course, was exactly the next giant step in the advancement of science.

2.5. *Series and computations*

In applying the approximation methodology to concrete situations, it is natural to denote the n-th approximate value by s_n. Thus $\{s_n\}$ constitutes a sequence converging to the target. It is often convenient to take the *incremental* approach by setting $a_n = s_n - s_{n-1}$, namely

$$s_n = a_1 + a_2 + \cdots + a_n. \tag{132}$$

Then, one has an infinite series $\sum_{i=1}^{\infty} a_i$ with s_n as the partial sum $\sum_{i=1}^{n} a_i$. Anyway, the technique of infinite series is a powerful tool in the application of approximation methodology. Among various types of infinite series, the power series and the Fourier series are the simpler and also more useful ones than other types, namely

$$\sum_{l=0}^{\infty} a_l x^l \tag{133}$$

$$\sum_{l=-\infty}^{\infty} c_l e^{ilx} = \sum_{l=-\infty}^{\infty} c_l(\cos lx + i\sin lx)$$

$$= c_0 + \sum_{l=1}^{\infty}(a_l \cos lx + b_l \sin lx) \tag{134}$$

where $a_l = c_l + c_{-l}$, $b_l = i(c_l - c_{-l})$.

We shall only include here a few specific examples of infinite series and their usages in specific computations. Let us begin with the following basic theorems on power series.

Theorem 5.8. *If there exist k and N such that*

$$\{|a_n|\}^{1/n} \le k \text{ for } n \ge N \tag{135}$$

then the power series

$$\sum_{l=0}^{\infty} a_l x^l$$

converges for all $|x| < \frac{1}{k}$.

Proof. Set $r = k \cdot |x| < 1$. Then, for $l \ge N$

$$|a_l x^l| = |a_l| \cdot |x|^l \le (k|x|)^l = r^l \tag{136}$$

and hence

$$\left|\sum_{l=m}^{n} a_l x^l\right| \le \sum_{l=m}^{n} |a_l x^l| \le \sum_{l=m}^{n} r^l \tag{137}$$

$$< r^m \cdot \frac{1}{1-r}$$

for all $m, n \ge N$.

Therefore, to any given $\varepsilon > 0$, there exists a sufficiently large N_ε such that

$$\left|\sum_{l=m}^{n} a_l x^l\right| \le \frac{r^m}{1-r} < \varepsilon \tag{138}$$

provided $m, n \geq N_\varepsilon$. (Notice that $0 < r < 1$ implies that $r^m \to 0$ as $m \to \infty$.) This proves the convergence of the power series by Cauchy's criterion. □

Theorem 5.9. *Let k and $\sum_{l=0}^{\infty} a_l x^l$ be the same as in Theorem 5.8 and set $f(x)$ as the limit value of $\sum_{l=0}^{\infty} a_l x^l$. Then*

$$F(x) = \int_0^x f(t)dt = \sum_{l=0}^{\infty} \frac{a_l}{l+1} x^{l+1}, \ |x| < \frac{1}{k}. \tag{139}$$

Proof. For any given $\varepsilon > 0$, there exists N_ε such that

$$\left| \sum_{l=N_\varepsilon}^{n} a_l x^l \right| < \frac{r^{N_\varepsilon}}{1-r} < \varepsilon \tag{140}$$

for all $n > N_\varepsilon$. Therefore

$$\left| F(x) - \sum_{l=0}^{N_\varepsilon - 1} \frac{a_l}{l+1} x^l \right| = \left| \int_0^x f(t)dt - \int_0^x \left(\sum_{l=0}^{N_\varepsilon - 1} a_l t^l \right) dt \right|$$

$$= \left| \int_0^x \left(\sum_{l=N_\varepsilon}^{\infty} a_l t^l \right) dt \right| \leq \varepsilon \cdot |x|. \tag{141}$$

This proves that

$$\sum_{l=0}^{\infty} \frac{a_l}{l+1} x^l = F(x). \tag{142}$$

□

Example 12 ($\tan^{-1} x$ **and the approximation of** π). Set $y = \tan^{-1} x$. Then $x = \tan y$. Hence, by the chain rule

$$1 = \frac{d}{dy} \tan y \cdot \frac{dy}{dx} = (1 + \tan^2 y)\frac{dy}{dx} = (1 + x^2)\frac{dy}{dx}. \tag{143}$$

This shows that

$$\frac{d}{dx} \tan^{-1} x = \frac{1}{1+x^2}. \tag{144}$$

For $|x| < 1$, it is easy to see that

$$\frac{1}{1+x^2} = 1 - x^2 + x^4 - \cdots + (-1)^n x^{2n} + \ldots . \tag{145}$$

Therefore, by Theorem 5.9

$$\tan^{-1} x = \int_0^x \frac{dt}{1+t^2} = x - \frac{x^3}{3} + \frac{x^5}{5} - \cdots + (-1)^n \frac{x^{2n+1}}{2n+1} + \cdots \tag{146}$$

for $|x| < 1$.

Recall that $\tan^{-1} 1 = \frac{\pi}{4}$ and

$$\tan^{-1} A + \tan^{-1} B = \tan^{-1} \frac{A+B}{1-AB}. \tag{147}$$

Hence

$$\begin{aligned}
\tan^{-1}\frac{1}{2} + \tan^{-1}\frac{1}{3} &= \tan^{-1} 1 = \frac{\pi}{4} \\
\tan^{-1}\frac{1}{3} + \tan^{-1}\frac{1}{7} &= \tan^{-1}\frac{1}{2} \\
\tan^{-1}\frac{1}{5} + \tan^{-1}\frac{1}{8} &= \tan^{-1}\frac{1}{3} \\
\frac{\pi}{4} &= 2\tan^{-1}\frac{1}{5} + 2\tan^{-1}\frac{1}{8} + \tan^{-1}\frac{1}{7}.
\end{aligned} \tag{148}$$

One needs to compute just a few terms of (146) in order to get highly accurate approximate values of $\tan^{-1}\frac{1}{5}$, $\tan^{-1}\frac{1}{7}$ and $\tan^{-1}\frac{1}{8}$, thus obtaining approximate value of π with a high order of precision.

Exercise. Try to compute π up to the tenth digit using the power series of $\tan^{-1} x$.

Example 13 (The computation of the logarithm table).

$$\frac{d}{dx} \ln(1+x) = \frac{1}{1+x} = 1 - x + x^2 - \cdots + (-1)^n x^n + \ldots \tag{149}$$

for $|x| < 1$. Hence, for $|x| < 1$, one has

$$\ln(1+x) = \int_0^x \frac{dt}{1+t} = x - \frac{x^2}{2} + \frac{x^3}{3} - \cdots + (-1)^n \frac{x^{n+1}}{n+1} + \ldots \quad (150)$$

In particular,

$$-\ln 2 = \ln\left(1 - \frac{1}{2}\right)$$

$$= -\left\{\frac{1}{2} + \frac{1}{2}\left(\frac{1}{2}\right)^2 + \frac{1}{3}\left(\frac{1}{2}\right)^3 + \cdots + \frac{1}{n}\left(\frac{1}{2}\right)^n + \ldots\right\}. \quad (151)$$

Exercise. Use the above formula to compute the approximate value of $\ln 2$ up to the 6th digit.

Next let us use the power series of $\frac{1}{2}\ln\frac{1+x}{1-x}$,

$$\frac{1}{2}\ln\frac{1+x}{1-x} = \frac{1}{2}\left\{\ln(1+x) - \ln(1-x)\right\}$$

$$= x + \frac{x^3}{3} + \frac{x^5}{x} + \cdots + \frac{x^{2n+1}}{2n+1} + \ldots \quad (152)$$

By substituting $x = \frac{1}{3}$ into the above formula, one gets

$$\frac{1}{2}\ln 2 = \frac{1}{3} + \left(\frac{1}{3}\right)^4 + \frac{1}{5}\left(\frac{1}{3}\right)^5 + \frac{1}{7}\left(\frac{1}{3}\right)^7 + \ldots \quad (153)$$

Exercise. Use the better formula above to compute $\ln 2$ up to the 12th digit.

For a prime number $p > 2$, set $x = \frac{1}{2p^2-1}$. Then

$$\frac{1+x}{1-x} = \frac{p^2}{p^2-1}$$

$$\frac{1}{2}\ln\frac{1+x}{1-x} = \frac{1}{2}\left\{2\ln p - \ln(p+1) - \ln(p-1)\right\}. \quad (154)$$

Hence

$$\ln p = \frac{1}{2}\ln(p+1) + \frac{1}{2}\ln(p-1)$$

$$+ \left\{ \frac{1}{2p^2-1} + \frac{1}{3}\left(\frac{1}{2p^2-1}\right)^3 \right. \tag{155}$$

$$\left. + \frac{1}{5}\left(\frac{1}{2p^2-1}\right)^5 + \frac{1}{7}\left(\frac{1}{2p^2-1}\right)^7 + \cdots \right\}.$$

Notice that both $(p+1)$ and $(p-1)$ can always be factored into the product of smaller primes. Thus the above formula effectively reduces the computation of $\ln p$ to that of the smaller primes and the rapidly convergent infinite series in the parenthesis. For example

$$\ln 3 = \left(1\frac{1}{2}\right)\ln 2 + \left\{ \frac{1}{17} + \frac{1}{3}\left(\frac{1}{17}\right)^3 + \frac{1}{5}\left(\frac{1}{17}\right)^5 + \cdots \right\}$$

$$\ln 5 = \frac{1}{2}\ln 3 + \left(1\frac{1}{2}\right)\ln 2 + \left\{ \frac{1}{49} + \frac{1}{3}\left(\frac{1}{49}\right)^3 + \frac{1}{5}\left(\frac{1}{49}\right)^5 + \cdots \right\}$$

$$\ln 7 = 2\ln 2 + \frac{1}{2}\ln 3 + \left\{ \frac{1}{97} + \frac{1}{3}\left(\frac{1}{97}\right)^3 + \frac{1}{5}\left(\frac{1}{97}\right)^5 + \cdots \right\}$$

$$\ln 11 = \left(1\frac{1}{2}\right)\ln 2 + \frac{1}{2}\ln 3 + \frac{1}{2}\ln 5 + \left\{ \frac{1}{241} + \frac{1}{3}\left(\frac{1}{241}\right)^3 + \cdots \right\}. \tag{156}$$

Exercise. Use the above formulas to compute $\ln 3$, $\ln 5$, $\ln 7$ and $\ln 11$ up to the tenth digit.

Example 14. Series of the type

$$\sum_{l=1}^{\infty} f(l) \tag{157}$$

where $f(x)$ is a *monotonically decreasing* function with $f(x) \to 0$ as $x \to \infty$, e.g.

$$
\begin{aligned}
f(x) &= \frac{1}{x} : 1 + \frac{1}{2} + \frac{1}{3} + \cdots + \frac{1}{n} + \cdots \\
f(x) &= \frac{1}{x^2} : 1 + \frac{1}{2^2} + \frac{1}{3^2} + \cdots + \frac{1}{n^2} + \cdots \\
f(x) &= \frac{1}{x^s} : 1 + \frac{1}{2^s} + \frac{1}{3^s} + \cdots + \frac{1}{n^s} + \cdots
\end{aligned}
\tag{158}
$$

The partial sum $\sum_{l=1}^{n} f(l)$ has a simple relationship with the integral of $f(x)$, namely, as indicated by Fig. 42,

$$
\sum_{l=1}^{n} f(l) \geq \int_{1}^{n+1} f(x)dx \geq \int \sum_{l=2}^{n+1} f(l).
\tag{159}
$$

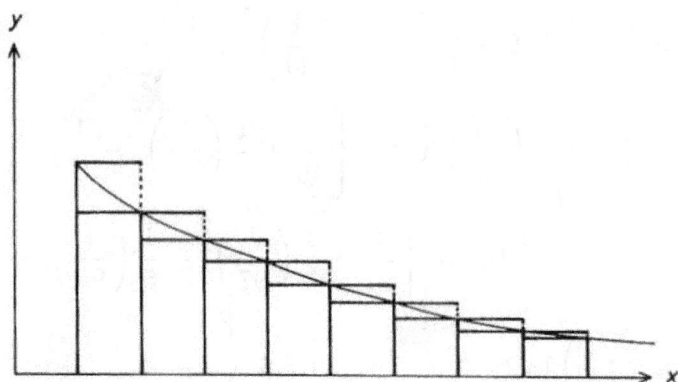

Fig. 48

Therefore, the infinite series $\sum_{l=1}^{\infty} f(l)$ converges if and only if $\int_{1}^{x} f(t)dt$ remains bounded as $x \to \infty$.

For example

(i) $1 + \frac{1}{2} + \frac{1}{3} + \cdots + \frac{1}{n} + \cdots$ diverges because

$$
\int_{1}^{x} \frac{1}{t}dt = \ln x \to \infty \text{ as } x \to \infty.
\tag{160}
$$

(ii) $1 + \frac{1}{2^s} + \frac{1}{3^s} + \cdots + \frac{1}{n^s} + \ldots$ converges for all $s > 1$ because

$$\int_1^x \frac{1}{t^s} dt = \frac{1}{s-1} \left(1 - \frac{1}{x^{(s-1)}} \right) \rightarrow \frac{1}{s-1} \qquad (161)$$

as $x \rightarrow \infty$.

Hence, the above infinite series defines a function for $s > 1$, namely

$$\zeta(s) = 1 + \frac{1}{2^s} + \frac{1}{3^s} + \cdots + \frac{1}{n^s} + \ldots \qquad (162)$$

This is the important ζ-*function* defined by Euler. The following identity of Euler demonstrates the fundamental importance of ζ-function in number theory, namely

Euler's Formula. $\zeta(s) = \prod_p \left(1 - \frac{1}{p^s} \right)^{-1}$, $s > 1$ where p runs through all prime numbers p.

Proof. For each given p, one has

$$\left(1 - \frac{1}{p^s} \right)^{-1} = 1 + \frac{1}{p^s} + \frac{1}{p^{2s}} + \frac{1}{p^{3s}} + \ldots \qquad (163)$$

Therefore

$$\prod_p \left(1 - \frac{1}{p^s} \right)^{-1} = \prod_p \left(\sum_{j=0}^{\infty} \frac{1}{p^{js}} \right)$$

$$= \sum_{n=1}^{\infty} \frac{1}{n^s} = \zeta(s) \qquad (164)$$

follows directly from the unique factorization of each n into the product of exponents of prime numbers, often referred as the fundamental theorem of arithmetic. In fact, the above Euler's formula, in one single formula, encompasses the totality of unique factorization of all natural numbers!

Using the basic property of the logarithmic function, one has

$$
\begin{aligned}
\ln \zeta(s) &= \sum_p \left(-\ln \left(1 - \frac{1}{p^s} \right) \right) \\
&= \sum_p \left(\sum_{j=1}^{\infty} \frac{1}{j} \left(\frac{1}{p^s} \right)^j \right) \\
&= \sum_{j=1}^{\infty} \left(\frac{1}{j} \sum_p \frac{1}{p^{js}} \right).
\end{aligned} \tag{165}
$$

Remark. Although the above formula only holds for $s > 1$, the limiting case of $p = 1$ still provides valuable information on the distribution of prime numbers, namely

$$
\sum_p \frac{1}{p} = \infty. \tag{166}
$$

Historically, it was already well-known that there are infinitely many prime numbers in antiquity. Euclid's Elements contained a beautiful proof of this fact. However, the divergence of $\sum \frac{1}{p}$ means the distribution of prime numbers *cannot* be very sparse! For example, it should be denser than the distribution of $\{n^s, \ n \in \mathbb{N}\}$ for any $s > 1$ because

$$
\sum_n \frac{1}{n^s} < \infty. \tag{167}
$$

Example 15 (π and e, the tale of two numbers). This book began with the discovery of irrationals by Hippasus and the Eudoxian methodology of approximation which enables us to understand the real number systems as a whole. We shall conclude this concise introduction of calculus by a brief discussion on two outstanding real numbers, namely, π and e, the former is the most important number naturally arises in the study of geometry while the latter is the most important number naturally arises in the study of analysis. In the setting of calculus, these two outstanding numbers are intimately linked by the following formula, namely

$$e^{i\pi} = -1. \tag{168}$$

Generally speaking, real numbers can be roughly classified into the following categories, namely
(i) *rational* numbers and *irrational* numbers.
(ii) *algebraic* numbers and *transcendental* numbers.

The ratios between the diagonal of a regular pentagon (resp. square) and its side, namely, $\frac{1}{2}(1 + \sqrt{5})$ (resp. $\sqrt{2}$), are irrational numbers.

However, they are the real roots of some polynomials with *integral coefficients*, namely,

$$(x^2 - x + 1) \text{ (resp. } x^2 - 2).$$

A number is defined to be algebraic if it can be realized as the root of a suitable polynomial with integral coefficients. Otherwise, it is called a transcendental number. π and e are both transcendental numbers. This remarkable fact was first proved for e by Hermite and then it was also proved for π by Lindermann. The key idea of Lindermann's proof of the transcendency of π is to make use of the relationship of $e^{i\pi} = -1$ to reduce the proof of transcendency of π to that of the proof of Hermite for the transcendency of e! This is a beautiful story of two outstanding numbers. We refer to books on transcendental number theory for those who are interested in learning this fascinating tale of two numbers.

www.ingramcontent.com/pod-product-compliance
Lightning Source LLC
Chambersburg PA
CBHW050642190326
41458CB00008B/2383